Ecocritical Geopolitics

What is the role of popular culture in shaping our discourse about the multi-faceted system of material things, subjects and causal agents that we call "environment"? *Ecocritical Geopolitics* offers a new theoretical perspective and approach to the analysis of environmental discourse in popular culture. It combines ecocriticial and critical geopolitical approaches to explore three main themes: dystopian visions, the relationship between the human, post-human, and "nature" and speciesism and carnism.

The importance of popular culture in the construction of geopolitical discourse is widely recognized. From ecocriticism, we also appreciate that literature, cinema, or theatre can offer a mirror of what the individual author wants to communicate about the relationship between the human being and what can be defined as non-human. This book provides an analysis of environmental discourses with the theoretical tools of critical geopolitics and the analytical methodology of ecocriticism. It develops and disseminates a new scientific approach, defined as "ecocritical geopolitics," to offer an idea of the power of popular culture in the realization of environmental discourse.

Referencing sources as diverse as *The Road*, *The Shape of Water*, *Lady and the Tramp*, and TV cooking shows, this book will be of great interest to students and scholars of geography, environmental studies, film studies, and environmental humanities.

Elena dell'Agnese teaches political geography and cultural geography at the University of Milano-Bicocca, where she is also Director of the Centre of Visual Research. Her work has been mainly focused on developing a wide-spectrum approach to "peripheral geographies." For this reason, she is interested in any form of — apparently innocent — "geo-graphical representation," from movies to television drama, cartoons and popular music, with specific attention given to issues relating to politics, gender, and race. She publishes extensively on these topics, mostly in Italian and in English, but also in French, Spanish, Japanese, and Croatian. In 2009, she founded the Association of Italian Geographers Study Group on "Media and Geography," which she chaired until 2015. She is now Vice-President of the Società Geografica Italiana. In 2014, she was elected Vice-President of the International Geographical Union.

Routledge Explorations in Environmental Studies

An Environmental History of Australian Rainforests until 1939
Fire, Rain, Settlers and Conservation
Warwick Frost

Daoism and Environmental Philosophy
Nourishing Life
Eric S. Nelson

Ecological Law and the Planetary Crisis
A Legal Guide for Harmony on Earth
Geoffrey Garver

From Environmental to Ecological Law
Kirsten Anker; Peter D Burdon; Geoffrey Garver; Michelle Maloney and Carla Sbert

Climate Change Temporalities
Explorations in Vernacular, Popular, and Scientific Discourse
Edited by Marit Ruge Bjærke, Anne Eriksen and Kyrre Kverndokk

Rights of Nature
A Re-examination
Edited by Daniel P. Corrigan and Markku Oksanen

Ecocritical Geopolitics
Popular Culture and Environmental Discourse
Elena dell'Agnese

Environmental Defenders
Deadly Struggles for Land and Territory
Edited by Mary Menton and Philippe Le Billon

For more information about this series, please visit: https://www.routledge.com/Routledge-Explorations-in-Environmental-Studies/book-series/REES

Ecocritical Geopolitics

Popular Culture and Environmental Discourse

Elena dell'Agnese

Routledge
Taylor & Francis Group
LONDON AND NEW YORK

First published 2021
by Routledge
2 Park Square, Milton Park, Abingdon, Oxon OX14 4RN

and by Routledge
605 Third Avenue, New York, NY 10158

Routledge is an imprint of the Taylor & Francis Group, an informa business

British Library Cataloguing-in-Publication Data
A catalogue record for this book is available from the British Library

Library of Congress Cataloging-in-Publication Data
Names: Dell'Agnese, E. (Elena), author.
Title: Ecocritical geopolitics : popular culture and environmental discourse / Elena dell'Agnese.
Description: Abingdon, Oxon ; New York, NY : Routledge, 2021. | Series: Routledge explorations in environmental studies | Includes bibliographical references and index.
Identifiers: LCCN 2020055338 (print) | LCCN 2020055339 (ebook) | ISBN 9780367264994 (hardback) | ISBN 9781032010748 (paperback) | ISBN 9780429293504 (ebook)
Subjects: LCSH: Geopolitics. | Ecocriticism. | Popular culture.
Classification: LCC JC319 .D44 2021 (print) | LCC JC319 (ebook) | DDC 304.2–dc23
LC record available at https://lccn.loc.gov/2020055338
LC ebook record available at https://lccn.loc.gov/2020055339

ISBN: 978-0-367-26499-4 (hbk)
ISBN: 978-1-032-01074-8 (pbk)
ISBN: 978-0-429-29350-4 (ebk)

Typeset in Bembo
by Newgen Publishing UK

Contents

Acknowledgments

This book would not exist if Rebecca Brennan, from Routledge publishing house, having seen an abstract of mine at the Association of American Geographers in 2016, had not suggested the idea. Rebecca is therefore the first person I have to thank.

This book would have been written in bad English, and full of errors, if Tim Stroud had not lost many of his days and probably nights reviewing and correcting the text. Tim is therefore the second person I want to thank.

The third person I thank is my son Edoardo Floriani. I know that family members are usually thanked at the end, for their support and patience. This case is different. Edoardo opened me to the discussion on the themes of veganism and carnism, and so I owe to him, and his ethical practice, many intellectual stimuli, which allowed me to develop the fourth part of the volume. I thank him also for having suggested many films and television series, and for having insisted that I see them all. And I thank my younger son Amedeo, for having the patience to watch them with me.

Then I would like to thank, for the bibliographical suggestions and advice, Marcella Schmidt of Friedberg, Virginie Mamadouh and Gerard Toal, in addition to the three anonymous reviewers. For the discussions Franco Farinelli, Fabio Amato, Bruno Vecchio, Alfonso Pecoraro Scanio, and Giorgio Mosterts; for the intellectual collaboration in the animal studies field, Ivan Bargna, Anna Mannucci and Gabi Scardi; for being a model to imitate, Michael Shapiro; for the appreciation they have shown towards me, giving me the confidence I need, Claude Raffestin, Vladimir Kolossov, and John O' Loughin; for the technical suggestions, Chiara Giubilaro and Stefano Malatesta; for the spur to act (and to write), Marco Grasso and Antonio Schizzerotto. I thank also the riders who in the months of the 2020 lockdown, when most of the Italian libraries and bookstores were closed, delivered me books otherwise untraceable, allowing me to work.

I thank Maria Chiara Zerbi for having taught me much of what I have learned. What I have not learned is not her fault. I would also like to thank Giacomo Corna Pellegrini, who is no longer here, for having taught me that intellectual freedom, in academic life, is the strongest weapon.

I would like to thank the following publishing houses for allowing the publication of excerpts from their titles: Penguin Random House, GROVE/ATLANTIC, INC., ECW Press Ltd., Simon & Schuster, P.O.L., Princeton University Press, Little, Brown and Company, an imprint of Hachette Book Group, Inc. (the data relating to the permissions are given in the notes, in connection with each excerpt). I would also like to thank Grace Harrison, Christine Bondira, Mary Dalton, and the other people who collaborate with Routledge for valuable support and help.

And finally, I thank all the members of my multispecies family, for keeping me company on or under my desk, bearing the deprivation of play and walks that I have imposed on them in these last two years.

Introduction

I.1 Why we need an "ecocritical geopolitics"

I.1.1 From Katniss Everdeen to Greta Thunberg (and back)

We all know who Greta Thunberg is. However, for those who are not too familiar with Young Adult Literature and the films adapted from it, it is necessary to explain who Katniss Everdeen is. When Greta became famous, she was a teenager. Like Greta, Katniss is a teenage girl who rebels against the system. She is the main character in a very famous literary trilogy (*The Hunger Games*, 2008–10, by Suzanne Collins) and its equally successful film adaptation (together, they form a "franchise"). According to the author's website,[1] over 100 million copies of the books have been sold, while the four films have grossed around three billion dollars worldwide.[2]

Katniss struggles against a system which is unjust economically, socially, and environmentally. To do so, she has to fight against a world of adults. Her heroic figure thus breaks gender, age, and class stereotypes. She defines herself as a fighter capable of promoting new ideals against corruption, vice, and the greed of previous generations. For this reason, she has been described as "emblematic of a new kind of progressive, female, Hollywood action hero" (Kirby 2015, 462).

The popularity of *The Hunger Games* has helped launch the fashion of fiction with a rebel girl as the protagonist. Katniss is arguably the most admired in what is now a long list of teenage heroines, from Hermione Granger in the *Harry Potter* saga, Beatrice Prior, called Tris, in the *Divergent* series, Rey in the *Star Wars* franchise, to a girl called Jiya in the Pakistani TV cartoon series *Burka Avenger* (Kimball 2019).

There is a long list of teenage heroines in the real world as well; some are famous worldwide, like Malala Yousafzai and Greta Thunberg, others only locally (Kimball 2017). As Gayle Kimball (2019) writes: "media empowers brave girls to be global activists." The question is how this happens and what kind of relationship there is between media heroines like Katniss and real ones like Greta. It is conceivable though not certain that one of the young "rebels" listed by Gayle Kimball has read a book or seen a film in the series *The Hunger*

Games, or another text starring a media heroine. It is much more likely that real figures inspired some of the media heroines. For instance, Jiya from *Burka Avenger* (2013–) struggles to defend the right for girls to go to school, just as Malala did (Pirzada 2017). She even broadens Malala's scope and promotes environmental awareness, like Katniss and Greta. However, "empowering media images" (Kimball 2019) does not necessarily imply that media images have a direct influence on inspiring human behavior, or that real people influence media images. More simply, the public's attention to media heroines in many cases paves the way for similar attention to be paid to the real rebels, giving them credit and the freedom to act. Alternatively, popular culture may create a milieu where the figure of a teenager fighting the system becomes acceptable For sure, the media convey (and in turn are influenced by) a particular way of seeing the world and our relations with that world, human and nonhuman.

Trying to figure out which is this "particular way of seeing the world and our relations with that world, human and nonhuman" is important, not only for analytical interest, but also, and perhaps above all, to understand if it is possible to intervene, if necessary, to modify it. In the theoretical framework of post-structuralism, the set of interpretative categories, values, and logics that we take for granted, and which correspond to this "particular way of seeing the world" is called "discourse," to use the term in the Foucauldian sense, which is to say as "the condition for the production and ordering of meaning and knowledge" (Rossini 2006). In order to understand our relation with the world, human and not human, the analysis must therefore lead to an understanding of the "discourse" about the "environment."

I.1.2 Popular culture and the "environment"

The analysis must put together the message, the genre, and the taken-for-granted world of the author to understand what in popular culture is going to filter through to its audience. Then, of course, there is the audience itself, which can interpret all these things in its own ways.

The first question to ask is "How do we define 'environment'?" The production of texts, whether filmic, literary, videogames or graphic novels, which deal with themes related to the "environment" is very wide. Even if the text we want to examine is presented as a "green" text (i.e., a product that conveys an environmentalist message), it is important to understand whether it presents "the environment" as "a juxtaposed externality in which humans dwell and conflict" in a dualistic approach, or instead as a "model that includes a wider number of subjects and causal agents, introducing issues of ontology, epistemology, and posthuman ethics" (Iovino 2012, 59–60). If the "environment" is dualistically considered a resource, the message will probably be a bit anthropocentric, and suggest that it must be conserved, because otherwise future generations will not have the same resources that we do. In the other case, it will be about the relative importance of humans among the multitude of other species living on the planet.

The genre and the taken-for-granted world of the author are also meaningful. Thus, a western movie may promote a conservationist message, displaying "a nostalgic yearning for the unspoiled wilderness," (Ingram 2000, 15) and at the same time belong to a film genre that exalts the alienation of a people from its land (the Native Americans), the near extinction of an animal species (the bison), and the construction of a regional economy based on the exploitation of an animal species for commercial purposes (cattle). A dystopian representation of the future may warn against the excessive power of corporations, consumerism, waste; at the same time, it may strengthen race and gender stereotypes, because in the taken-for-granted world of the author in question women are only housewives and people of color subordinates. Alternatively, a film about intensive farming and the American meat industry may at the same time mourn the end of traditional extensive farming, falling into the dichotomy typical of carnism, between the meat that one should not eat, because it comes from industrial farms and meat that one can eat since it comes from extensive farms.

The context where the action takes place also matters. Almost every text, even if it does not have a deliberately environmental message, has a setting, and therefore it speaks in some way of the system of material things, subjects, and causal agents that we call "environment." If we reflect on the relationship between human beings and whatever is perceived as "other" (nonhuman animals, plant life, other humans) the range widens further. Is popular culture still communicating dualisms, separating the "human" *par excellence* (male, white, handsome, healthy), from those who are not male, not beautiful, not healthy, not white, or is it making an effort to overcome these dichotomies with a different approach?

In this regard, everyday actions performed by characters may count. Thus, questions multiply: is it "normal" to see a fictional program on TV where the family has bacon and eggs for breakfast, or should we consider it a proof of the strength of the taken-for-granted exceptionality of humans as a species? Usually, it is considered "normal" to eat animals of certain species but not others, for instance, not the ones we keep as pets. Speaking of pets, there are many books and movies about dogs. How do we represent the relationship we have with the animals we say we love? Is it dominance or affection (Tuan 1984)? Lastly, who can speak? The omniscient narrator, perhaps from a presumably objective point of view, or the "subaltern"? Is it possible to give a voice to "our dumb animals" (as suggested by the title of a US magazine dedicated to animal welfare)? If so, how, without falling into anthropomorphism? And finally, we can investigate whether writing is more effective than images, albeit violent and graphic, in preventing animals from being sent to slaughter.

I.1.3 Ecocritical geopolitics

All these questions may be formulated from different scientific perspectives. For this reason, the theme of the representation of "environmental" issues by popular culture is addressed by various research approaches, from media studies, to English literature, cultural studies and environmental cultural studies (Sturgeon 2009), but also sociology, ecocriticism and environmental humanities (Opperman and Iovino 2016). So why do we need one more?

The proposal advanced here is aimed at emphasizing the connection between power and knowledge, and, in particular, the power of representation over reality. To do this, it is suggested that the theoretical frame developed by critical geopolitics should be used. As described by McFarlane and Hay (2003, 212), critical geopolitics problematizes "prevailing geopolitical orders not as given, but as 'historically constructed perspectives asserting privileged forms of representation'" and "challenges the unremarkable quality of hegemonic representations within the international sphere."

Popular geopolitics, a subfield of critical geopolitics, specifically "demands that attention be given to examination of the role of the media in the construction and perpetuation of dominant geopolitical understandings" (McFarlane and Hay 2003, 211). The idea is also that the way we interpret the "environment" is connected with a somewhat hegemonic representation that deserves to be challenged, and that this representation is perpetuated through popular culture.

"We are interested in how popular culture constructs and reveals spatial and political fields of meaning," state Robert A. Saunders and Vlad Strukov (2018), introducing a reader to the subject (*Popular Geopolitics: Plotting an Evolving Interdiscipline*). To these "spatial and political fields of meaning" we wish to add environmental ones. From a methodological point of view, this approach will combine the analytical tools of geography with those of ecocriticism, while theoretically it will be heavily indebted to other approaches, such as ecofeminism, posthumanism, and environmental cultural studies, which seek to bring together environmental issues with reasoning on the relationship between human beings (Sturgeon 2009). I propose the term "ecocritical geopolitics."

The Hunger Games provides excellent examples from this point of view. Katniss' role as a teenage heroine has aroused considerable scholarly interest about the saga in terms of gender, class, race, and also environment. In "an ecocritical reading," Janice Bland and Anne Strotmann (2014) identify the literary tropes related to environmental issues in the book trilogy. The tropes are the "apocalypse," because the setting of the saga is in a post-apocalyptic future in North America; the "pastoral," which creates the spatial distinction of city and countryside and the temporal distinction of a dreadful present and an idyllic past; and the "wilderness," the natural one surrounding the district where Katniss lives, and the artificial one in the arena of the Capitol, where she has to prove her bravery. Wilderness is not just the only space where freedom is to be had, it also plays an important role in the story, since Katniss manages to survive in the Capitol arena precisely because of her wilderness skills (even though, in order to acquire them, she broke the law by going hunting across the district boundary).

This also enables a look at the trilogy through an ecofeminist lens. Katniss' closeness to nature leads Tan (2017) to propose a parallel between her rebellion against patriarchy and nature's revolt against the industrial system: "As a result, Katniss' quest in *The Hunger Games* is homogenous to nature's quest for survival against an industry-driven totalitarian regime which exploits nature and its resources to the fullest extent. Nonetheless, the survival and rebellion of Katniss is equally symbolical for the revival and continuation of nature" (Tan 2016, 39).

With a similarly ecofeminist approach, Burke (2013) stresses how the trilogy brings to the fore the issue of environmental justice. The post-apocalyptic world where Katniss lives is represented as a geopolitical system where the rich city center exploits the poor suburbs environmentally to produce its food, while the locals starve. To get food, Katniss, who lives in one of the poorest districts, learns to hunt (breaking the law). In this way, she fights against environmental injustice.

The themes of feminist rebelliousness and socio-spatial injustice are also taken up by Kirby in an article in the journal *Geopolitics* (Kirby 2015), where he tries to systematize a theoretical approach to critical geopolitics that includes the suggestions of feminist geography and paying attention to audience response (in order to avoid the risk of "abstracting popular texts from context, place and reception," Kirby 2015, 464). In the paper, Kirby emphasizes Katniss' positive reach and her great success on social media and feminist blogs; yet he notes that the trilogy falls into several clichés, like "the classic philosophical position, popularised by Henry Thoreau, that goodness comes from a proximity to nature" (Kirby 2015, 476).

Overall, the analysis carried out by Kirby remains anchored to the themes of critical geopolitics (socio-spatial inequality, class, race), even if in part it conjugates them with those of feminist geography (gender). An ecocritical geopolitics approach could similarly underline that Katniss fights against socio-spatial and gender inequalities, and that her struggle launches a message about environmental justice; however, beyond the cliché that Kirby underlines, attention should also be drawn to the fact that Katniss' attitude towards the environment remains utilitarian (she goes hunting and, after admiring its beauty, kills a deer) (Burke 2015) and therefore that the discourse developed by the trilogy is still anchored to mainstream anthropocentric assumptions. In addition, an approach linked to geographical analysis tools could help highlight how wilderness is represented as opposed to humanized space, in a rather stereotyped, and dualistic, reading of territorial diversity.

I.1.4 Chapters outline and structure

The first part of this book, dedicated to introducing critical geopolitics, popular geopolitics, ecocriticism and their research tools, will be followed by three sections of "applied" analysis. Each part is divided in chapters, each chapter in sections. The three topics analyzed are chosen by way of example, to see how ecocritical geopolitics can be used to analyze a narrative strategy — such as dystopian and post-apocalyptic narration — a theme, the posthuman, and a discourse, carnism.

The three areas of analysis have been chosen for their relevance. The apocalypse is considered the environmental metaphor *par excellence*. Dystopias and post-apocalyptic narratives tell of human beings' fears, and therefore of their relationship with their environment. In addition, they are often accompanied by environmental and landscape descriptions capable of offering an insight from this perspective.

The posthuman theme is the theoretical approach that best helps us to reflect on the limits of the definition of human and nonhuman, and more generally on the dualisms typical of the Western tradition. If dystopian and post-apocalyptic themes reflect fears about the future of humanity, a posthumanist approach allows us to think about the representation of our relationship with technology, on the one hand, and with nonhuman animals, on the other.

Lastly, if "landscape epitomizes aesthetic conceptualizations of nature," food is one of the principal modes of negotiating and representing the nature – culture relationship, because it turns nature into "a consumable thing" (Gersdorf 2010, 4). "Carnism" is a way of considering the relationship between human and nonhuman animals, which, on the basis of a high degree of speciesism, takes eating dead animals for granted. Eating meat is "normal," "natural" and "necessary" (Joy 2010), even if cattle farming is today one of the main environmental threats globally. According to a 2006 FAO report, "Expansion of livestock production is a key factor in deforestation"; moreover, in terms of climate change, livestock "is a major player," responsible for a percentage of greenhouse gas emissions measured in CO_2 equivalent, which is higher than transport. It is also "a key player in increasing water use, [...] mostly for the irrigation of feedcrops. It is probably the largest sectoral source of water pollution, contributing to eutrophication, 'dead' zones in coastal areas, degradation of coral reefs, human health problems, emergence of antibiotic resistance and many others." It should also be added that livestock is also a threat to biodiversity, "since it is the major driver of deforestation, as well as one of the leading drivers of land degradation, pollution, climate change, overfishing, sedimentation of coastal areas and facilitation of invasions by alien species. In addition, resource conflicts with pastoralists threaten species of wild predators and also protected areas close to pastures" (Steinfeld et al. 2006).

Deconstructing carnism as discourse therefore has not only an ethical purpose, with regard to nonhuman animals, but also a political sense, in terms of environmental conservation.

Each of the three areas of analysis is also characterized by the prevalence of a different environmental discourse, albeit with notable exceptions. Dystopian narratives tend to express a mainstream concern about environmental issues, conveying a conservationist attitude towards the environment as a resource and a predominantly anthropocentric mindset, although they often address social issues and can include important reflections on environmental justice.

Post-apocalyptic fiction can overcome the anthropocentrism of dystopia, and offer a biocentric point of view, even if it can hardly overcome the stereotypes of race and gender typical of dystopian fiction. The posthuman often involves an ecofeminist approach, and carnism may be questioned, to challenge anthropocentrism and speciesist ideologies.

Analyzing a narrative genre for its environmental discourse requires that a great number of texts are considered, examined more for their clichés and points in common than for their artistic uniqueness. For this reason, Chapter 4 examines the birth and evolution of dystopian and post-apocalyptic narrative strategies, the relationship between fear and popular culture, and the role of dystopian and post-apocalyptic narratives, as a vehicle to express the fears of their times. It then highlights what fears have been expressed over time by these narrative strategies and how environmental issues have become so mainstream in recent years as to be the object of dystopian parodies. Eventually, it is stressed that the great popularity of post-apocalyptic and dystopian narratives risks undermining their communicative power.

Chapter 5 focuses on the use of the setting and landscape in dystopian and post-apocalyptic literature. After proposing a distinction between the two — the setting is the description of the context, while the landscape presumes a relational dimension between the human being and the context — the chapter examines settings and landscapes of dystopian and post-apocalyptic narratives. As far as dystopian narratives are concerned, since they are generally anthropocentric in approach, the setting is heavily artificial, generally characterized by a high urban density and a marked territorialization. Nature, within a fairly traditional dualism, is opposed to culture. It exists externally, beyond a border, and constitutes an element of disorder, but also of freedom, to aspire to (or as a possible place to flee to). These green spaces of freedom are a recurrent element in dystopic (and sometimes post-apocalyptic) fiction, even if, as other texts make clear, a Planet B is not yet conceivable. In addition to the border between anthropic and natural spaces, dystopian territorializations may also include a spatial demarcation of social inequalities. The setting is therefore characterized by rigid boundaries between the spaces of the privileged and those of the poor, spaces which, in texts straddling dystopia and science fiction, can also extend to artificial worlds. In the spaces of the poor, the representation of conditions of environmental exploitation and pollution also triggers issues of environmental justice. The fear of the overwhelming power of capitalism and corporations, the conditioning of consumers and the excessive consumerism that follows is increasingly present in dystopian narratives. The recurring metaphor in this regard is that of the accumulation of waste. Some post-apocalyptic narratives give the setting an even more important value, or even exploit the relational dimension between the human protagonist and what surrounds him/her to charge the landscape with a metaphorical dimension. In this regard, the chapter examines three post-apocalyptic texts, which, in a biocentric perspective, deal with a future in which the planet continues its history, indifferent to the disappearance of the human species. On the contrary, it becomes more luxuriant and richer than other forms of life. The texts are *The Last Man* (1826, Mary Shelley), *Earth Abides* (1949, George Stewart) and *The Drowned World* (1962, J.G. Ballard). The chapter ends with the analysis of a text that has enjoyed great critical acclaim: *The Road* (2006, Cormac McCarthy). Here too the landscape has great importance. In this case, however, the biosphere was destroyed, and the planet died. Taking a more anthropocentric approach than the previous texts, Cormac McCarthy focuses on the spiritual dimension of the human being, and the ability to "carry the fire" (i.e. love), even when the struggle for resources is extreme.

The main characters in all the texts examined in Chapter 5 are males, white and non-disabled. The next chapter, Chapter 6, therefore examines how race and gender stereotypes enter into dystopian and post-apocalyptic narratives and how, and whether, they have any connection with environmental discourse. The "heroic male agent" is the first object of attention, as this figure reflects all the possible factors for discrimination: those of gender (he is male and heterosexual), race (he is white), age (he is usually in his 30s), and those related to the body (he is always, or almost always, without disabilities). This, together with the fact that he is often also a father, allows him to have a protective attitude not only towards women and the family, but also towards nature. This anthropocentric dualism is also reflected in the more recent narrative, where adolescent and rebellious heroines have emerged (as women, they are closer to nature, according to the old stereotype). In addition to human heroines, the chapter also examines virtual, android, or digital female characters, to show how, even in this case, there is often an overlapping domination (over women and technology). Finally, the theme of the apocalypse deserves to be read from the point of view of those who have already had such an experience, that is, have seen their world end, to be replaced by a different world. Native Americans and many indigenous peoples have already experienced their apocalypse. In this regard, I consider *Moon of the Crusted Snow* (2018), a post-apocalyptic novel by the Anishinaabe author Waubgeshig Rice. The novel, which plays with the tropes of mainstream post-apocalyptic literature, shows how those fears are something external to a people who have already experienced their own apocalypse.

In Part 3, I turn to a philosophical approach, which may be present in narratives of different genres, posthumanism. Chapter 7 starts with the quote: "Posthuman is a seductive term, being both nebulous and popular" (Gray 2017, 148), then seeks to shed light on three expressions that are sometimes used in overlapping ways: post-human, transhuman, and posthuman. We use the expression post-human, with the hyphen, to define something that comes after the human, in relation to something that has to do with evolution, or rather with the possible future evolution of the human being. The term "posthuman," not hyphenated, is used to indicate the desire (or need) to overcome the binary distinction between the human and the "other," whatever the other is. Transhuman is related to a technologically enhanced human being. Posthumanism considers the posthuman product of transhumanism (the cyborg) as a metaphor (Haraway 1985), in order to break three dichotomies — between human and animal, between organism and machine, and between physical and nonphysical — and to overcome all the binary oppositions typical of Western society, such as male/female, nature/culture, human/animal, and so on. Having set this theoretical framework, the chapter moves on by first analyzing post-human scenarios, like those proposed in the novel *The Time Machine* (1895, H.G. Wells), then the loss of humanity, or most of the salient features of humanity, in different texts presenting a variation of zombie apocalypse, from the novel *I Am Legend* (1954, Richard Matheson) to the videogame *The Last of Us*; control over fertility and genetic manipulation, with the goal of improving the human species, in the texts *Brave New World* (1932, Aldous Huxley) and the *MaddAddam* trilogy (2003–13, Margaret Atwood); the review is concluded with transhuman creations, in the form of androids and cyborgs.

The topic of the cyborg introduces the theme of the posthuman, which is analyzed in relation to popular culture in Chapters 8 and 9. To investigate posthuman and transhuman perspectives in popular culture, science fiction and dystopian narratives provide an excellent platform. The posthuman approach is broader, perhaps too broad to be fully explored within these frameworks. So, to limit the scope of research, a first step in this direction has been taken by trying to understand how the relationship between humans and nonhuman animals is represented in terms of "companion species," as suggested by Donna Haraway (2003). To this end, it was decided to examine the representation of a nonhuman animal that has always had a particularly strong relationship with human beings, the dog (Chapter 8). A brief review of how the dog is represented in popular culture highlights how dogs are often anthropomorphized, gendered or stereotyped, and portrayed in an unbalanced relationship of power, in which they are items of property more than companions. The chapter goes on to show how the dog–human rapport, far from being based on companionship, risks being naturalized as a servant–master relationship with a connection symbolized by the leash and the joy that, in the "happy ending" of many dog stories, the dog is supposed to feel when it is given a collar. The process of denaturalizing the dog and imposing rules on him is present in many dog stories set in the American West (*The Old Yeller*, 1956; *Stickeen*, 1897; *White Fang*, 1906; *Where the North Begins*, 1923), but also in the city, as evidenced by the famous animated cartoon *Lady and the Tramp* (1955). In these dog stories, the relationship between nature and culture is still seen in the rigid perspective of binarism, with nature that must be dominated, and the human being as the dominator (the holder of the leash).

Those who lead a dog on a leash may believe that they love the dog, even if in fact they dominate it. On the contrary, for those who practice vivisection, there is no doubt. What is interesting to note is that even among those who oppose vivisection the motivations are often anthropocentric (Section 8.3). Even more marked is the anthropocentrism at the basis of the practice of vivisection, but among antivivisectionists attitudes can also vary between strongly anthropocentric positions (vivisection has the effect of making practitioners insensitive) and more biocentric/posthuman attitudes (the extreme diversity of power between the animal victim, totally deprived of agency, and the human executioner). Section 8.4 is dedicated to certain dog stories that express antivivisectionist feelings, and to the antivivisection debate. The subject is very interesting from the point of view of ecocritical geopolitics, on the one hand because the debate has involved some of the greatest exponents of the culture of the time (such as Lewis Carroll, the poets Tennyson and Browning, and the authors H.G. Wells and G.B. Shaw), who mobilized popular culture for the cause; on the other hand because the movement was marked by a large presence of women activists and intellectuals, sharing with feminism the fight against objectification, body control, and male domination. In the section, I consider some poetic texts (*Tray*, 1879), novels and short stories (*A Dog's Tale*; 1903; *Trixy*, 1904) where the canine protagonist risks, or even ends up on the vivisector's table.

The chapter then examines three texts that differ from most dog stories, precisely because of their ability to deal with the relationship between human beings and nonhuman animals from a posthuman perspective. The first is *Flush: a Biography* (1933) by Virginia Woolf, in which the writer, with a great exercise of imagination, manages to represent the "smellscape" of the dog and to penetrate his *Umwelt* (Uexküll 1934). In this way, the reader is introduced to the spatial experiences of a dog in a way that challenges the empirical belief in the authority of vision. The second novel is *A Dog of Flanders* (1872) by Ouida; the book is a curious literary case since it is very famous in Japan but almost forgotten in Europe, albeit its author is English. The novel is about the friendship of a human being (a boy, named Nello) and a dog (Patrasche) that describes a very intense and equal relationship, and illustrates arguments similar to those later developed by Donna Haraway (2003). The third novel is *Animal's People* (2007), a text by a contemporary Anglo-Indian author, Indra Shina. Again, the story is about the friendship between a boy, who is disabled and walks on all fours, and a dog named Jara. The story takes place in a fictional city that stands for Bhopal in the aftermath of the gas tragedy, a setting that allows the author to raise a series of fundamental themes. Beyond the posthuman approach connected with the relationship between the boy (called Animal) and Jara the dog, the novel touches on the themes of "slow violence," linked to the environmental damage caused by the gas tragedy, and environmental justice.

Chapter 9 goes one step further in the direction of posthumanism when it addresses the notion of the "hybrid." Hybrids, "monsters," and "extra-ordinary bodies" provide the means to consider the themes of dualism, dichotomizations, and easy classifications. After an introduction to the different notions of monstrous, from classical culture onwards, a review of the representation of the non-"normate" body is made, with the idea that "normate" refers to "the constructed identity of those who, by way of the bodily configurations and cultural capital they assume, can step into a position of authority and wield the power it grants them" (Garland Thompson 2017, xii). We thus reach the position of posthumanism, which internalizes the hybrid, the "cyborg," the monster, as a starting point, to destabilize the symbolic boundaries set by the notion of the human, as referred to in Western history. Such representations of fantastic animals and extraordinary beasts, bodily instability and mutations of various kinds can be read in the *Harry Potter* saga. In the second section, we examine two very different texts (a novel and a film) that share points in common. The novel is *The Dolphin People* (2006, Torsten Krol), and the movie *The Shape of Water* (2017, Guillermo del Toro). Their point of contact is given by the opposition between those who want to impose order, and therefore consider scientifically impure and expendable those who have an extra-ordinary body, and those who defend the right to be non-"normate." In both texts, the villain is impersonated by a character who demonstrates all the physical characteristics of the "human" imposed by Western humanism: he is a white male, heterosexual, handsome, and able-bodied. He represents "modernity." In both cases, the mysterious Otherness, in spatial term, is provided by the image of Amazonia. So, the two texts openly question body dualisms, but fail to overcome geographical ones, and fall into the spatial stereotype that what is outside the West is less civilized, less ordered, and closer to nature.

The third section (9.3) is devoted to the analysis of *Truismes* (1996, Marie Darrieussecq), a novel about a metamorphosis: the body (in this case, the body of a woman) here varies, in the course of the narration, from "normal" (indeed, beautiful) to "monstrous," because, through various stages, it becomes the body of a sow. Appreciated for its strong message against the exploitation of women, the novel also opens up a reflection that departs from traditional anthropocentric assumptions. Metamorphosis is a metaphor, used to highlight the violations that specifically animalize the narrator, but it is also a way to explore animality as a condition and different perception of the world, introducing the idea of hybridity as a way to overcome the separation between human and nonhuman animals. Finally, the novel leads inside a slaughterhouse, opening a "fissure" in carnism as a discourse.

One difficult question runs through almost all the texts examined in this part of the book: "Can the subaltern speak?" (Spivak 1988), that is, how can we interpret, and express, the sensitivity of nonhuman animals? Or, more generally, who is on their side? Stylistic responses vary, from attempting to write the autobiography of an animal, to assuming animals' point of view, to simply offering an "*écriture de cochon*" (literally, "pig writing," that is, writing in a messy way), as the narrator-sow of *Truismes* refers to her writing.

The fourth part deals with a commonsensical attitude: carnism. Carnism, as defined by Melanie Joy (2010), is the dominant discourse of human nutrition; based on rigidly speciesist assumptions, carnism considers eating meat to be normal, natural, and necessary. This part of the book is dedicated to the deconstruction of carnism as a discourse. The question here is even more challenging: is it possible to produce a popular culture that is capable of going against the dominant ideology? Does it make sense to defend animals in writing? Or is it better to replace writing with visual evidence? The answer to the latter case can only be given by the audience.

After introducing "carnism," Chapter 10.1 presents the "Three Ns of the justification" (normal, natural, and necessary), and the three cognitive mechanisms that lead to a distortion of perceived reality, which Joy (2010) calls "the Cognitive Trio" ("objectification," "deindividualization," and "dichotomization"). The chapter then introduces the "meat paradox" represented by the awareness that eating animals involves killing them, the conflict triggered by this awareness between our desire to be moral people and our desire to eat meat, and the role played by advertising when it skirts around the fact that meat-eating requires animal deaths. Lastly, the role of literature, cinema, and television in replicating the dominant way of viewing animals and food is highlighted, together with the "carnonormativity" (Freeman 2014) of popular culture. There are few vegetarians or vegans; moreover, as a brief review shows, if they are men they are "different," if not, they are girls.

The association between meat consumption (and, more generally, nonhuman animal domination) and gender is explored in Chapter 11, where the concepts of "carno-fallogocentrism" (Derrida 1991) and "anthroparchy" (Cudworth 2005) are introduced, then the connection between masculinity and meat in cultural traditions and in media is explored. A narrative genre that glorifies the combination "masculinity–meat" is the western movie, which celebrates the breeding of wild cattle as one of the founding elements of American national identity, and its workers (the cowboys) as national heroes. At the same time, the genre manages to glorify the genocide of a people (Native Americans), the almost total extinction of an animal species (the bison), and the exploitation of another (cattle). It succeeds in bringing together anthroparchy, racism, and a geographical dualism in the land of "the others" (the West), which is conceived as virgin land to be conquered. Another form of popular culture that promotes animal-based food, meat, and masculinity is represented by culinary talent shows on TV. Particular attention is paid to *Masterchef*, the most successful culinary talent show, where the judges are almost always male chefs. This kind of show not only contributes to normalizing and even promoting the consumption of meat and other animal products, but helps to create the figure of the chef as a media star, fostering a gender discourse, where masculinity implies power and hierarchical superiority. In this way, once again, carnism is mixed with gender issues and popular culture.

Chapter 12 changes perspective and discusses whether popular culture can undermine the dominant discourse. In particular, through the analytical approach of Stam, Burgoyne, and Flitterman-Lewis (2005), I try to highlight which products of popular culture can be classified as "resistant," "formally resistant," and "fissure" texts, with reference to carnism as ideology and also for problematizing representations of nonhuman animals and consumption practices. In the first part of the chapter, feature films for adults (a few) and children (more), which somehow question carnism and anthroparchy, are examined. I then examine some texts that, by mixing fiction and investigative journalism, have partly succeeded, in both past and present, in disturbing readers' consciences: *The Jungle* (1906, Upton Sinclair), *Fast Food Nation* (2001, Eric Schlosser) and *Eating Animals* (2009, Jonathan Safran Foer). Later, attention is turned to literary texts written by critically acclaimed writers with the same objective: to unhinge carnism. The texts taken into consideration are *The Lives of Animals* (1999) by J.M. Coetzee, *The Slaughterer* (1967) by Isaac Bashevis Singer, and *Consider the Lobster* (2004) by David Foster Wallace. Though with different styles, the three texts embrace empathy rather than denunciation, addressing the core tactics of carnist discourse, such as dichotomization or denial. A mention is also made of novels that, though dedicated to other events, open, with a single sentence or a paragraph, a crack in the carnist discourse (such as *Benoni*, 1908, by Knut Hamsun).

Finally, taking inspiration from a novel entitled *My Year of Meats* (1998, Ruth Ozeki), the chapter questions the role of materials of visual denunciation, which have sometimes been considered too violent to achieve the hoped-for result, and yet seem to have converted some viewers to vegetarianism and veganism.

Notes

1 www.suzannecollinsbooks.com/bio.htm
2 www.the-numbers.com/movies/franchise/Hunger-Games#tab=summary

Bibliography

Bland, Janice, & Strotmann, Anne (2014). The Hunger Games: An Ecocritical Reading. *Children's Literature in English Language Education* 2(1), 22–43.

Burke, Brianna R. (2013). Teaching Environmental Justice Through The Hunger Games. *The ALAN Review* 43(1), 53–63. https://doi.org/10.21061/alan.v41i1.a.7

Burke, Brianna R. (2015). "Reaping" Environmental Justice Through Compassion in The Hunger Games. *ISLE: Interdisciplinary Studies in Literature and Environment* 22(3), 544–567. https://doi.org/10.1093/isle/isu099

Cudworth, Erika (2005). *Developing Ecofeminist Theory: The Complexity of Difference.* Palgrave Macmillan.

Derrida, Jacques (1991). Eating Well, or the Calculation of the Subject: An Interview with Jacques Derrida. In Cadava, Eduardo, Connor, Peter, & Nancy, Jean-Luc (eds.). *Who Comes After The Subject?* (pp. 96–118). Routledge.

Freeman, Carrie P. (2014). Lisa and Phoebe, Lone Vegetarian Icons: At Odds with Television's Carnonormativity. In Macey, Deborah A., Ryan, Kathleen M., & Springer, Noah J. (eds.), *How Television Shapes Our Worldview: Media Representations of Social Trends and Change* (pp. 193–212). Lexington Books.

Garland Thompson, Rosemary (2017). Introduction: From Wonder to Error. A Genealogy of Freak Discourse in Modernity. In Garland Thompson, Rosemary (ed.), *Freakery: Cultural Spectacles of the Extraordinary Body* (pp. 1–19). New York University Press.

Gersdorf, Catrin (2010). Authenticity Redux: Ecology and the 'Ethics of Representing' in Ruth Ozeki's *My Year of Meat.* In Haselstein, Ulla, Gross, Andrew & Snyder-Körber,

Maryann (eds.), *The Pathos of Authenticity: American Passions of the Real* (pp. 59–73). Universitätsverlag Winter.

Gray, Chris Hables (2017). Post-Sapiens: Notes on the Politics of Future Human Terminology. *Journal of Posthuman Studies* 1(2), 136–150. www.jstor.org/stable/10.5325/jpoststud.1.2.0136

Haraway, Donna (1985). Manifesto for Cyborgs: Science, Technology, and Socialist Feminism in the 1980s. *Socialist Review* 80, 65–108 (also as A Cyborg Manifesto: Science, Technology, and Socialist-Feminism in the Late Twentieth Century. In Haraway, Donna (ed.), *Simians, Cyborgs, and Women*, Routledge, 1991, and in Haraway, Donna, *Manifestly Haraway*, University of Minnesota Press, 2016).

Haraway, Donna (2003). *The Companion Species Manifesto: Dogs, People, and Significant Otherness*, Prickly Paradigm Press (also in Haraway, Donna. *Manifestly Haraway*, University of Minnesota Press, 2016).

Ingram, David (2000). *Green Screen. Environmentalism and Hollywood Cinema*. University of Exeter Press.

Iovino, Serenella (2012). Material Ecocriticism: Matter, Text, and Posthuman Ethics. In Müller, Timo, & Sauter, Michael (eds.), *Literature, Ecology, Ethics*. (pp. 51–68). Universitätsverlag Winter.

Joy, Melanie (2010). *Why We Love Dogs, Eat Pigs, and Wear Cows: An Introduction to Carnism*. Conari Press.

Kimball, Gayle (2017). *Brave: Young Women's Global Revolution*. Equality Press.

Kimball, Gayle (2019). Media Empowers Brave Girls to be Global Activists. *Journal of International Women's Studies* 20(7), 35–56.

Kirby, Philip (2015). The Girl on Fire: The Hunger Games, Feminist Geopolitics and the Contemporary Female Action Hero. *Geopolitics* 20(2), 460–478. https://doi.org/10.1080/14650045.2014.984835

McFarlane, Thomas, & Hay, Iain (2003). The Battle for Seattle: Protest and Popular Geopolitics in The Australian Newspaper. *Political Geography* 22(2), 211–232. https://doi.org/10.1016/S0962-6298(02)00090-2

Oppermann, Serpil, & Iovino, Serenella (2016). *Environmental Humanities: Voices from the Anthropocene*. Rowman & Littlefield International.

Pirzada, Tehmina (2017). Narrating Muslim Girlhood in the Pakistani Cityscape of Graphic Narratives. *Girlhood Studies* 10(3), 88–104.

Rossini, Manuela (2006). To the Dogs: Companion Speciesism and the New Feminist Materialism. *Kritikos* 3, 1–25

Saunders, Robert A., & Strukov, Vlad (eds.) (2018) *Popular Geopolitics: Plotting an Evolving Interdiscipline*. Routledge.

Spivak, Gayatri Chakravorty (1988). *Can the Subaltern Speak?* Macmillan.

Stam, Robert, Burgoyne, Robert & Flitterman-Lewis, Sandy (2005). *New Vocabularies in Film Semiotics*. Routledge.

Steinfeld, Henning, Gerber, Pierre, Wassenaar, Tom D., Castel, Vincent., Rosales, Mauricio, & de Haan, Cees (2006). *Livestock's Long Shadow: Environmental Issues and Options*. Food & Agriculture Org.

Sturgeon, Noël (2009). *Environmentalism in Popular Culture: Gender, Race, Sexuality, and the Politics of the Natural*. University of Arizona Press.

Tan, Cenk (2017). Rebellious Women in Men's Dystopia: Katniss and Furiosa. *Pamukkale University Journal of Social Sciences Institute* 26, 32–46. http://acikerisim.pau.edu.tr:8080/xmlui/handle/11499/26037

Tuan, Yi-Fu (1984). *Dominance and Affection*. Yale University Press.
von Uexküll, Jakob (1934) Streifzüge durch die Umwelten von Tieren und Menschen: Ein Bilderbuch unsichtbarer Welten. J. Springer (2010, *A Foray into the Worlds of Animals and Humans: With a Theory of Meaning*. University of Minnesota Press).

1

Theoretical framework

1 Geo(-)graphy, critical geopolitics, popular geopolitics

1.1 "Geo-graphy is about power"

The connection between representation and reality is intrinsic to the very name of the discipline. The word "geo-graphy" comes from the Latin word *geōgraphia*, which in turn comes from the Ancient Greek γεωγραφία (geōgraphía), from γῆ (gê), land or earth, and γράφω (gráphō), to draw or to write. "Geography" means "writing" or "drawing" the Earth. At the same time, the term "geography" is currently used to refer to the material object, the Earth, with its regional characteristics and differences (see Robert Kaplan's book, *The Revenge of Geography: What the Map Tells Us About Coming Conflicts and the Battle against Fate*, 2012).

This terminological ambiguity, according to which both the object and the representation of the object are defined with the same word, postulates, according to Claude Raffestin (1983), the perfect adaptation between the object of discourse and the discourse. So, geo-graphy, as a discourse, is placed outside historical conditioning, while, far from being a-historical, it is controlled by power. And power, through geographical discourse, manages to make the geo-structure it produces more acceptable. Representation, which is supposed to be the same as reality, but in fact is historically determined and influenced by power, makes reality "normal" and therefore "acceptable."

Before Raffestin, already Yves Lacoste (1976) in France, and Lucio Gambi (1968) and Massimo Quaini (1974) in Italy had shown how knowledge of the territory, and therefore geography, was an instrument at the service of the prince, to administer, to collect taxes, and to make war. Quaini showed that geography is an instrument of power in a dual sense: it supplies useful information to the ruling classes while simultaneously providing mystifying information to the subaltern classes, thus negating them the chance of a proper knowledge of space and power.

To this path of reasoning, however, Raffestin adds something more. Geography is not just an instrument in the hands of powerful people, it is power in itself, because geography is representation and representation has the power to change reality. This suggestion, clearly connected with Foucault's idea of power-knowledge (Raffestin 1978), has since been developed in other — but somehow interconnected — contexts of European and American academic thought.

Critical attention to geography as representation, and its power to transform reality, was also urged by Giuseppe Dematteis (Fall and Minca 2013), who, in 1985, published, in Italian, *Le metafore della Terra* (The Metaphors of the Earth). Here, he takes up the idea of geography as an "active" practice, capable of modifying reality. He writes that: "The ideological function of 'textbook geography' is [...] complex [...] Not only does it make geographical knowledge look like an innocent form of knowledge, and not only does it teach people that what exists is *natural* and *cannot* be changed, but also, and more subtly, it teaches people that what exists is *normal* [...] that it is natural *because it is normal*" (Dematteis 1985, 10).

Then, he adds that "geographical representation, based on evidence and appealing to common sense, produces con-sense [...] from which normalized behaviors derive, that is, behaviors that can be integrated into collective practices. These, acting on Earth, transform it into territory, which in turn is the object of geographical representation" (1985, 101).

Geography represents the world and produces consensus, but it is not a specular reproduction of the world. It is just a set of metaphors. Only the awareness that these metaphors are just partial representations of reality can help people to escape, according to Dematteis, from the loop of their "normalizing" effects (dell'Agnese 2008).

In geographical representation, not only information provided by writing or statistics matters. Cartographic visualization counts as a form of power as well. In this regard, Franco Farinelli's work opens a new page in the analysis of geography as a form of geo-power. Farinelli distinguishes geo-writing and geo-mapping, and their different roles in transforming places into spaces. Mapping means trying to represent a sphere on a flat surface. To do so, one must move away from the sphere, i.e., put one's perspective above reality, and assume that one can read/dominate the world from above. From the presumption of zenithal vision comes what Farinelli himself defines as "cartographic reason" (Farinelli 1992 and Farinelli 1998), an attitude typical of Western thought, which leads to seeing the world in two dimensions, as if it were a "table." Thinking in cartographic terms, that is along with cartographic reason, means detaching oneself from reality, and presuming to be able to read/dominate it from above. The notion of "cartographic reason," as developed by Franco Farinelli (1992), but also by Gunnar Olsson (1998) and Tom Conley (1996), can be seen as "the missing element in social theories of modernity" (Pickles 2004, xi), as it helps to emphasize "the role of mapping in shaping social, spatial and natural identities" (Pickles 2004, xi). The role of maps in the connection between power and knowledge has also been pointed out by Harley (1988) who, drawing on Foucault's ideas, defines maps as perfect instruments to promote a hierarchical vision of the space they represent.

1.2 Critical geopolitics/popular geopolitics

The awareness of the power of geography to produce and transform the world by describing it is one of the most important developments in contemporary geographical thought. Hence, when Gerard Toal opens his book *Critical Geopolitics* (Ó Tuathail 1996, 1) with the sentence "Geography is about power," he is not inventing something new. However, he is formalizing a combination of ideas in a clear theoretical approach and on that approach, he bases a new way of doing and thinking geopolitics. He writes: "Although often assumed to be innocent, the geography of the world is not a product of nature but a product of histories of struggle between competing authorities over the power to organize, occupy, and administer space [...]. Geography was not something already possessed by the earth but an active writing of the earth by an expanding, centralizing imperial state. It was not a noun but a verb, a geo-graphing, an earth-writing" (Ó Tuathail 1996, 1).

In this way, he restates the idea that "the land is not only represented, it is also produced by geography." Starting from Gerald Toal's contribution, Foucault's notions of power/knowledge and discourse spark the revolution in geography advocated by Raffestin (1997). Since then, the act of "writing the world" has no longer been seen as a straightforward performance. On the contrary, the idea that writing, representation and mapping are all forms of geo-power is now mainstream among the practitioners of the discipline.

Critical Geopolitics has been defined as "[i]maginative, intellectually ambitious [...] engaging [...] outstanding" (Hague 2011, 417). For this reason, the book has become a reference text for political geography worldwide. However, it is not the first step in this direction. Starting from the conviction that geography is never a non-discursive phenomenon, separated from ideology and politics, but a form of knowledge in the Foucauldian sense, Toal wrote, together with John Agnew, an article in 1992 where they proposed "the reconceptualization of geopolitics using the concept of discourse" (Ó Tuathail and Agnew 1992). In the paper, geopolitics is defined as "a discursive practice by which intellectuals of statecraft 'spatialize' international politics and represent it as a 'world' characterized by particular types of places, peoples and dramas" (Ó Tuathail and Agnew 1992, 190). At the beginning of the 1990s, Simon Dalby was also moving in the same direction with the book *Creating the Second Cold War* (Dalby 1990) and the article "Critical Geopolitics: Discourse, Difference, and Dissent" (Dalby 1991).

The notion of discourse refers to the interpretative, not simply verbal, tools that are put in place for the construction of meaning. In Foucauldian terms "discourses are not simply reflections or (mis)representations of 'reality'; rather they create their own 'regimes of truth'— the acceptable formulation of problems and solutions to those problems" (Lees 2004, 102–103). In this perspective, the geopolitical is "a differential discourse that mobilizes spatial logics and imaginaries for processes of identity construction, while presenting itself as merely reporting on the final product of these processes" (Grayson 2018). A geopolitical discourse is always "localized" in time and space and has a "naturalizing" function, i.e., it allows us to accept as "natural" things that are not natural at all. Hence, it must be deconstructed starting from its rhetorical artifices, the reference lexicon, the basic postulates and all those categories which, although they are themselves a product of a given historical and cultural context, are instead considered bearers of an interpretative capacity of absolute value. In this regard, any form of representation that has as its object the depiction of spaces and/or spatially connoted individuals constitutes a form of knowledge/power capable of producing a discursive narrative and deserves to become an object of research, including popular culture.

A geopolitical discourse is also embedded, i.e., deeply rooted in historical, geographical and cultural reality. At times, a certain tradition of thought, and the ensuing discourse, takes precedence over the others and prevails. In this perspective can be read, for example, the imposition of the Westphalian system of the nation-state on all other potential forms of political-spatial control of power. Within this "territorial trap" (Agnew 1994), the dominant geopolitical discourse can be defined as a process of spatial exclusion: "it can be argued that the essential moment of geopolitical discourse is the division of space into 'our' place and 'their' place" (Dalby 1991, 274), dividing "our" national territory, but also "our" urban neighborhood, from "their" national territory and "their" urban district. Geopolitical discourse not only works by dividing spaces, it also divides human beings associated with them, "its political function being to incorporate and regulate 'us' or 'the same' by distinguishing 'us' from 'them', the same from 'the other'" (ibid.). According to this theoretical approach, "geopolitics is about the assignment of values to places, and it constructs hierarchies of people and places that matter and those that do not" (Dittmer and Bos 2019, xix).

To understand how a given geopolitical discourse develops, that is which values are assigned to people and places, it is important to analyze representations. They can be produced from a wide variety of sources. In this regard, Gerald Toal (Ó Tuathail 1999) distinguishes between different types of representation: those provided by "formal geopolitics," i.e., the set of studies and research that form the object of geopolitics as a discipline; those consisting of the discourses and practices of active politics, which are part of so-called "practical geopolitics"; those produced by popular culture ("popular geopolitics"). Practical geopolitics, formal geopolitics, and popular geopolitics are all forms of power-knowledge, which tend to intersect. Practical geopolitics can influence popular geopolitics (through the production of propaganda films, for example), and be influenced by it, because "the distinction between the truly factual and fictional blurs in this society of the media spectacle" (Sharp 1998, 155), while "The mass media produce geo-graphs of world politics and international relations for public consumption alongside the more erudite highbrow texts mentioned above" (ibid., 153). Just as the production of discourse is embedded, i.e. deeply rooted in historical, geographical, and cultural reality, so is the audience (Dittmer and Dodds

2008), which will be able to negotiate different meanings from the same text, depending on its location in the world, on its competence, on its education.

Popular geopolitics is a subdiscipline developed within critical geopolitics (Dittmer 2018). Specifically, popular geopolitics studies how popular culture constructs people and places within broader political narratives and "focuses on the ways in which popular culture discourses contribute to the creation of hegemony, whether that is patriarchy (gender-based hegemony), bigotry (race-based hegemony), heteronormativity (sexual preference-based hegemony), or other forms of hegemony" (Dittmer and Bos 2019, 30). Cinema has been popular geopolitics' main object of interest for a long time, since movies provide "a language and imagery," but also "reference points and ways of enframing popular understandings of the […] world" (Power and Crampton 2005, 193). Nowadays, much of the attention of popular geopolitics' practitioners still focuses mostly on visual media: art, cinema, comics, video games. Occasionally, it focuses on music (Kirby 2019) and song lyrics (dell'Agnese 2015). Sometimes, it also works on message boards and social media (Harby 2019). More rarely, it takes into account literary texts (novels, short stories, poems).

Starting from its origins (the expression "popular geopolitics" was first used by Joanne Sharp in 1993), popular geopolitics has broadened its interests so much that it is very difficult to conduct an exhaustive analysis of all the trends that are being implemented (Saunders and Strukov 2018). At the same time, it has established itself as an indispensable approach to geopolitical analysis, aimed at revealing how certain representations, ideas and forms of knowledge become rooted as common sense (Grayson 2018).

1.3 Ecocriticism

Widening the scope of analysis to include the study of the relationship between human and nonhuman brings critical geopolitics in contact with ecocriticism. Ecocriticism, as defined by Garrard (2009, 19), is "the ability to investigate cultural artifacts from an ecological perspective." It is a cultural criticism approach, developed in the 1990s to study the "relationships between […] human culture and the physical world" (Glotfelty 1996, xx).

Initially, there was no agreement on how to name it. Garrard (2010, 1) writes in this regard that "as leaders in the field, Lawrence Buell and Jonathan Bate have expressed a preference for the names 'environmental criticism' and 'ecopoetics.'" "Other suggested alternatives to 'ecocriticism' have included literary-environmental studies, literary ecology, literary environmentalism, and green cultural studies" (Hutchings 2007, 174). However, as Garrard (2010, 1) adds, ecocriticism now "is the most prevalent and widely accepted name."

The choice of name is important, also from a theoretical point of view. According to Cheryll Glotfelty, one of the scholars who has contributed most to give a canonical and recognizable form to the field, it is better to use the prefix "eco-," instead of building a portmanteau with the word environment. "In its connotations, enviro- is anthropocentric and dualistic, implying that we humans are at the center, surrounded by everything that is not us, the environment. Eco-, in contrast, implies interdependent communities, integrated systems, and strong connections among constituent parts" (1996, xix). Ecocriticism, in this way, puts itself outside the anthropocentric discourse.

Ecocriticism arises as a reaction to the awareness of the looming environmental crisis, with a clear political question: "How then can we contribute to environmental restoration, not just in our spare time, but from within our capacity as professors of literature?" Thus, just as feminist literary criticism shares with feminism the commitment to women, and Marxist criticism sides with the workers, so ecocriticism is committed to the environment (Glotfelty 1996, xxi). "As a field of literary inquiry [...] 'ecocriticism' [...] investigates literature in relation to the histories of ecological or environmentalist thought, ethics, and activism" (Hutchings 2007, 172). Thus, it aims to analyze the meaning attributed to the environment and nature by a particular author in a single text or by an entire literary genre, or to evaluate the evolution of concepts, such as "'nature', 'wilderness', 'humanity', 'the animal' and 'progress'" (Kerridge 2016, 14). Specifically, it aims to answer questions such as "What role does the physical setting play in the plot of this novel? Are the values expressed in this play consistent with ecological wisdom? How do our metaphors of the land influence the way we treat it? How can we characterize nature writing as a genre? In addition to race, class, and gender, should place become a new critical category? [...] What view of nature informs US Government reports, corporate advertising, and televised nature documentaries, and to what rhetorical effect?" (Glotfelty 1996, xix).

Ecocriticism goes beyond nature writing and the writing alone. Precisely because of the themes dealt with,ecocriticism promises to leave the rooms of literary criticism, to encourage the cross-fertilization of literary studies with other disciplines such as history, psychology, art history, ethics, and above all philosophy, especially with regard to the subject of posthumanism and ecofeminism (Iovino 2010; Oppermann 2013; Oppermann 2016). Along this path, ecocriticism even challenges "logocentric thought," theorizing a "material ecocriticism" approach, which merges "material realities into discursive dynamics" (Iovino and Opperman 2012, 448).

Thanks to these ongoing cross-fertilizations, the thematic interests of ecocriticism have greatly expanded over time. The very idea of "physical environment" has become immensely enriched, so much so that Greg Garrard, in a theoretical synthesis (2004, 5), extends the scope of analysis to include "the study of the relationship between human and nonhuman, throughout the cultural history of humanity, and the critical analysis of the very meaning of the term 'human'."

Parallel to the themes, the range of texts subject to analysis has also been extended, because a message about the environment may be "everywhere, from TV cookery programs to new readings of Shakespeare" (Kerridge 2016). So, from the initial core, focused on Anglo-American nature writing or English romantic poetry, the interest of ecocriticism has gone in the direction of different popular culture products, such as theatre performances, visual art, photography, cinema (Armbruster and Wallace 2001), but also "matter as a text, as a site of narrativity, a storied matter, a corporeal palimpsest in which stories are inscribed" (Iovino and Opperman 2012, 451).

From a methodological point of view, the analysis promoted by ecocriticism can offer many inspiring examples. As far as the analysis is concerned, ecocriticism suggests going beyond the simple examination of plot and characters to verify the use of metaphors, similes and other figures of speech related to the world of nature, or the quality of environmental descriptions and the kind of setting.

Given the attention expressed by ecocriticism towards topics such as environment and nature, sense of place and landscape, it seems difficult not to see a convergence of interests with geography, and in particular with those forms of geographical research that deal with analyzing the modes of representation of the "geo" (i.e., terrestrial spaces) present outside the texts of conventional geography. Moreover, as summarized by Serenella Iovino in an interview,[1] the relationship between ecology and literature can be considered taking either of two approaches: thematic and "systemic." The thematic approach highlights how nature and the environment are represented in cultural texts and how these representations have changed over time. This type of analysis not only deals with representations of natural disasters, or texts with an environmentalist message, because any text representing contact between humans and nonhumans (animals, context, natural life forms, or artificial intelligence) can be the subject of analysis. Then there is the approach, which Iovino calls "systemic." It does not focus on the ideas represented in the texts, but on how ideas interact with the socio-cultural contexts where it circulates. Thus, she says, it is possible to look at the role of literature and cultural texts in the collective imagination (in the hope that they can be instruments of "environmental education") (see also Iovino 2013).

The attention to the "effects" that popular culture (be it literary works, films, plays) can have in the way of thinking about the "environment" and, more generally, the relationship between humans and nonhuman elements of the context, is common to ecocritical geopolitics. However, even if the texts analyzed may coincide and the interest in environmental issues coalesce, between critical ecocriticism and the "expanded" version of geopolitics, which is "ecocritical geopolitics," there is still a difference. Ecocritical geopolitics goes beyond the formal analysis of the text, to seek the power-knowledge mechanism in its discursive articulation; moreover, it is also interested in seeing how the audience reacts.

Note

1 https://www.greenious.it/serenella-iovino-ecologia-letteratura-un-binomio-inscindibile/, consulted on 14/06/2020

Bibliography

Agnew, John A (1994). The Territorial Trap: the Geographical Assumptions of International Relations Theory. *Review of International Political Economy* 1(1), 53–80. https://doi.org/10.1080/09692299408434268

Armbruster, Karla, & Wallace, Kathleen R. (eds.) (2001). *Beyond Nature Writing: Expanding the Boundaries of Ecocriticism*. University of Virginia Press.

Conley, Tom (1996). *The Self-Made Map: Cartographic Writing in Early Modern France*. University of Minnesota Press.

Dalby, Simon (1990). *Creating the Second Cold War: The Discourse of Politics*. Bloomsbury Publishing,

Dalby, Simon (1991). Critical Geopolitics: Discourse, Difference, and Dissent. *Environment and Planning D: Society and Space* 9(3), 261–283. https://doi.org/10.1068/d090261

dell'Agnese, Elena (2008). Geo-graphing: Writing worlds. In Cox, Kevin R., Low, Murray, & Robinson, Jennifer (eds.), *The Sage Handbook of Political Geography* (pp. 439–453). Sage.

dell'Agnese, Elena (2015). "Welcome to Tijuana": Popular Music on the US-Mexico Border. *Geopolitics* 20(1), 171–192. https://doi.org/10.1080/14650045.2014.979914

Dematteis, Giuseppe (1985). *Le metafore della terra: la geografia umana tra mito e scienza*. Feltrinelli.

Dittmer, Jason (2018). The Origins And Evolution of Popular Geopolitics: An Interview with Jo Sharp and Klaus Dodds. In Saunders, Robert A., & Strukov, Vlad (eds.), *Popular Geopolitics: Plotting an Evolving Interdiscipline* (pp. 23–42). Routledge.

Dittmer, Jason, & Bos, Daniel (2019). *Popular Culture, Geopolitics, and Identity*. Rowman & Littlefield.

Dittmer, Jason, & Dodds, Klaus (2008). Popular Geopolitics Past and Future: Fandom, Identities and Audiences. *Geopolitics* 13(3), 437–457. https://doi.org/10.1080/14650040802203687

Fall, Juliet J., & Minca, Claudio (2013). Not a Geography of What Doesn't Exist, but a Counter-Geography of What Does: Rereading Giuseppe Dematteis' Le Metafore della Terra. *Progress in Human Geography* 37(4), 542–563. https://doi.org/10.1177/0309132512463622

Farinelli, Franco (1992). *I segni del mondo. Immagine cartografica e discorso geografico in età moderna*.La nuova Italia.

Farinelli, Franco (1998). Did Anaximander Ever Say (or Write) Any Words? The Nature of Cartographical Reason. *Ethics, Place & Environment* 1(2), 135–144, [http://doi.org/10.1080/1366879X.1998.11644223]

Gambi, Lucio (1968). *Geografia e contestazione*. Fratelli Lega.

Garrard, Greg (2004). *Ecocriticism*. Routledge.

Garrard, Greg (2009). Ecocriticism: The Ability to Investigate Cultural Artefacts from an Ecological Perspective. In Stibbe, Alan *The Handbook of Sustainability Literacy: Skills for a Changing World* (pp. 19–24). Green Books.

Garrard, Greg (2010). Ecocriticism. *Year's Work in Critical and Cultural Theory* 18(1), 1–35. https://doi.org/10.1093/ywcct/mbq005

Glotfelty, Cheryll (1996). Introduction: Literary Studies in an Age of Environmental Crisis. in Glotfelty, Cheryll, & Fromm, Harold (eds.), *The Ecocriticism Reader: Landmarks in Literary Ecology* (pp. xv–xxxvii). University of Georgia.

Grayson, Kyle (2018). Popular Geopolitics and Popular Culture in World Politics: Pasts, Presents, Futures. In Saunders, Robert A., & Strukov, Vlad (eds.), *Popular Geopolitics: Plotting an Evolving Interdiscipline* (pp. 43–62). Routledge.

Hague, Euan (2011). Gearóid Ó Tuathail (Gerard Toal). In Hubbard, Phil, Kitchin, Rob, & Valentine, Gill (eds.), *Key Thinkers on Space and Place*, 2nd ed. (pp. 226–230). Sage.

Harby, Alexander J. (2019). Historicising Popular Geopolitics. *Geography Compass* 13(1) e12416 https://doi.org/10.1111/gec3.12416

Harley, J. Brian (1988). Maps, Knowledge and Power. In Cosgrove, Denis, & Daniels, Stephen (eds.), *The Iconography of Landscape* (pp. 277–312). Cambridge University Press.

Hutchings, Kevin (2007). Ecocriticism in British Romantic Studies. *Literature Compass* 4(1), 172–202. [http://doi: 10.1111/j.1741-4113.2006.00417

Iovino, Serenella (2010). Ecocriticism and a Non-Anthropocentric Humanism: Reflections on Local Natures and Global Responsibilities. In Volkmann, Laurenz, Grimm, Nancy, Detmers, Ines, & Thomson, Katrin (eds.), *Local Natures, Global Responsibilities, Ecocritical Perspectives on the New English Literatures* (pp. 29–53). Brill Rodopi.

Iovino, Serenella (2013). Ecocritica: teoria e pratica. In Salabè, Caterina (ed.), *Ecocritica. La letteratura e la crisi del pianeta* (pp. 17–26). Donzelli.

Iovino, Serenella, & Oppermann, Serpil (2012). Theorizing Material Ecocriticism: A Diptych. *ISLE: Interdisciplinary Studies in Literature and Environment* 19(3), 448–475.

Kaplan, Robert (2012). *The Revenge of Geography: What the Map Tells Us About Coming Conflicts and the Battle against Fate.* Random House.

Kerridge, Richard (2016). Ecocriticism and the Mission of 'English'. In Garrard, Greg (ed.), *Teaching Ecocriticism and Green Cultural Studies* (pp. 11–23). Springer.

Kirby, Philip (2019). Sound and Fury? Film Score and the Geopolitics of Instrumental Music. *Political Geography* 75 102054 https://doi.org/10.1016/j.polgeo.2019.102054

Lacoste, Yves (1976). *La Géographie ça sert d'abord à faire la guerre.* Maspero.

Lees, Loretta (2004). Urban Geography: Discourse Analysis and Urban Research. *Progress in Human Geography* 28(1), 101–7. https://doi.org/10.1191/0309132504ph473pr

Ó Tuathail, Gearóid (1996). *Critical Geopolitics: The Politics of Writing Global Space.* University of Minnesota Press.

Ó Tuathail, Gearóid (1999). Understanding Critical Geopolitics: Geopolitics and Risk Society. *The Journal of Strategic Studies* 22(2–3), 107–124.

Ó Tuathail, Gearóid, Agnew, John (1992). Geopolitics and Discourse. Practical Geopolitical Reasoning in American Foreign Policy. *Political Geography* 11(2), 190–204. https://doi.org/10.1016/0962-6298(92)90048-X

Olsson, Gunnar (1998). Towards a Critique of Cartographic Reason. *Ethics, Place & Environment* 1(2), 145–155. 10.1080/1366879X.1998.11644224

Oppermann, Serpil (2013). Feminist Ecocriticism: A Posthumanist Direction in Ecocritical Trajectory. In Gaard, Greta, Estok, Simon C., & Oppermann, Serpil (eds.), *International Perspectives in Feminist Ecocriticism* (pp. 19–36). Routledge.

Oppermann, Serpil (2016). From Posthumanism to Posthuman Ecocriticism. *Relations Beyond Anthropocentrism* 4(1), 23–37. https://www.ledonline.it/index.php/Relations/article/view/990

Pickles, John (2004). *A History of Spaces: Cartographic Reason, Mapping, and the Geo-Coded World.* Psychology Press.

Power, Marcus, & Crampton, Andrew (2005). Reel Geopolitics: Cinemato-Graphing Political Space. *Geopolitics* 10(2), 193–203. https://doi.org/10.1080/14650040590946494

Quaini, Massimo (1974). *Marxismo e geografia*. La Nuova Italia (1982, *Marxism and Geography*. Basil Blackwell).

Raffestin, Claude (1978). Evoluzione storica della territorialità in Svizzera. In Racine, Jean Bertrand, Raffestin, Claude, & Ruffy, Victor (eds.), *Territorialità e paradigma Centro-Periferia*. Unicopli.

Raffestin, Claude (ed.) (1983). *Geografia Politica: teorie per un progetto sociale*. Unicopli.

Raffestin, Claude (1997). Foucault aurait-il pu révolutionner la géographie? In Rotmann, Roger (ed.), *Au risque de Foucault* (pp. 141–149). Éditions du Centre Pompidou. https://archive-ouverte.unige.ch/unige:4462

Saunders, Robert A., & Strukov, Vlad (eds.) (2018). *Popular Geopolitics: Plotting an Evolving Interdiscipline*. Routledge.

Sharp, Joanne P. (1993). Publishing American Identity: Popular Geopolitics, Myth and the Reader's Digest. *Political Geography* 12(6), 491–503. https://doi.org/10.1016/0962-6298(93)90001-N

Sharp, Joanne P. (1998). Reel Geographies of the New World Order: Patriotism, Masculinity, and Geopolitics in Post-Cold War American Movies. In Tuathail, Gearóid Ó., & Dalby, Simon (eds.), *Rethinking Geopolitics* (pp. 164–181). Routledge.

2 What kind of environmental discourse is that?

2.1 Discourse about the multifaceted system of material things, subjects and causal agents that may be called "environment"

A fundamental thing, when dealing with the analysis of a text using the approach of popular geopolitics, is to distinguish between the message, if there is one, and the discourse. Discourse is defined by Brulle as "the term used to describe the historically specific world-views that serve as the basis for formulation of collective social action" (1996, 60). If discourse is the set of taken-for-granted categories used to make sense of the world, the message can only be influenced by it. However, the message depends on a conscious choice by the author, while the discourse is linked to the cultural context in which the author is placed. Since it is taken for granted, discourse is what the author considers natural and normal, without questioning it. A green movie or a nature-writing book can therefore convey a message inspired by the author's desire to protect the environment and at the same time be imbued with a discourse in which the human being is given a role of exceptionality in relation to the other components of the environment. In fact, this is what happens most often. Moreover, as has already been said before, any other text can reflect a discourse on the environment, or rather on the relation, more or less "dualistic," that is supposed to exist between human beings and "nature," even without the desire to convey an environmental message.

Defining what is meant by environmental discourse is not easy, since both the term discourse and the term environment can be subject to a multiplicity of interpretations and uses. According to Giblett (2011, 9–10), talking about the "environment" is already a discourse above "nature." If nature can be "defined simply as a collective noun for land, living beings, air, water, energy and planetary motion," speaking about "environment" means separating "a subject from its environs," and producing a "master-slave relationship between them." In a less drastic way, Iovino does not propose to abandon the notion of environment but to define it as a "model that includes a wider number of subjects and causal agents, introducing issues of ontology, epistemology, and posthuman ethics" (Iovino 2012, 59–60).

About discourse, it may be intended merely in connection with Foucault's philosophy, as a linguistic and semiotic construction, or as a practice. Generally speaking, it can be assumed that "Discourse about the natural world, like discourse generally, is composed of fragments of multiple social texts, ideologies, or styles" (Marafiote and Plec 2011, 49). More specifically, it may be defined as the "dominant ideological premises that both precondition and reproduce particular human relationships with nature" (Milstein 2009, 25), no matter what form these premises take. The connection between discourse and material reality is in any case evident. As stated by Giblett (2011, 10), "One of the most powerful ways in which mastery over the earth is exercised is via the discourses of nature. I define discourses as institutionalized ways of seeing, saying and doing." According to Milstein (2009), "The exploration of the role of discourse in human destruction of the environment and oppression of other species is of central concern because, even though human power over nature is decidedly materially experienced, the material practices themselves are both justified and reinforced via processes and systems of discourse, or representation" (2009, 26)

Discursive practices are also power practices. Quoting Milstein again, it is possible to observe that "discursive practices that incorporate significations of nature, animals, and certain people as subordinated 'others' will inevitably be loaded with ideology that contributes to reinforcing the structure of power relations." (Milstein 2009, 26).

Alongside hegemonic, or dominant, discourse and its practices, it is possible that alternative discursive practices may develop, in contrast to them. Milstein lists, in this regard, three dialectics typical of Western discourse on the environment: first, the presumption that the human being has mastery over nature, and, in contrast, the idea that the ability to live in harmony with nature should be positively evaluated; second, there is the dialectic between reading difference in dichotomous terms, and the desire to overcome these dualisms ("This othering often serves not only to justify exploitive views and practices, but also to divorce humans from the knowledge that they are, in fact, animals and part of nature themselves", 2009, 27). Eventually, the third dialectic, which descends directly from the first two, opposes exploitation to idealism. It should be noted that mastering, othering, and exploitation are connected to an underlying attitude, anthropocentrism, which has its roots in ancient times. In addition, it must be emphasized, again with Milstein (2009, 28), that "Contemporary, profit-driven Western processes of excessive mass consumption largely depend upon mastery, othering, and exploitation of nature, animals, and other people." For this, they are much more pervasive than alternative discursive practices. Both will be examined in the following sections, even if, while being aware of the great contribution that oriental philosophies, or indigenous attitudes towards nature, could offer to the debate, the analysis will be limited to Western thought alone.

2.2 Anthropocentrism and speciesism

The most obvious of attitudes towards one's "environs" is to consider oneself at the center. Indeed, "some degree of an anthropocentric orientation is inescapable simply by the fact that *as* humans, humans perceive the world within the limits of human bodies and cognition" (Weitzenfeld and Joy 2014, 4). Indeed, "we are restricted to our own human interpretations of nature and can consider our treatment of the natural world only within the framework of human desires and needs" (Taylor 1983, 239).

Albeit "there are some ways where anthropocentrism is not objectable" (Hayward 1997, 49), anthropocentrism, which places humans at the center of all meaning, is not just a perceptive issue, but an attitude that takes on features that are ontological (human beings are exceptional) and ethical (and thus they are the masters of their universe and can exploit every other living being).

Anthropocentrism is so widespread as to appear "normal", or rather, "natural" and therefore normal; however, it is not just an innate disposition, it is "an historical outcome" (Weitzenfeld and Joy 2014) and has its own cultural specificity. It is typical of Western thought, for instance, but absent in many indigenous societies (Kopnina, Washington, Taylor and Piccolo 2018). The bio-regional home-habitat of the Australian Aboriginal Country, for instance, is composed of "symbiotic livelihoods [...] that work (with) the earth as a living being" (Giblett 2011, 10).

As far as Western thought goes, the idea has ancient origins and refers both to philosophy (Aristotle denied reason to animals, while the Stoics denied duties of justice toward them) and religion. Specifically, "The dominant view in the history of Western philosophy is that human beings are fundamentally superior to nonhuman animals, typically on the grounds that only human beings possess reason, language, and self-awareness" (Steiner 2005, 38). This attitude belongs to the whole "monolithic anthropocentric Aristotelian-Ptolemaic cosmology which came to dominate the Islamic and later the medieval European world-view" (Sessions 1974, 75). The view that nonhuman nature has no intrinsic claim to exist and to prosper in itself was later reinforced by the development of modern Western philosophy. The apex of this position was reached by René Descartes (1596–1650), according to whom not only do animals not have rights, but they do not even have sensitivity ("animals are without feeling or awareness of any kind"), because they are simple mechanisms (their body being "comme une machine qui, ayant été faite des mains de Dieu, est incomparablement mieux ordonnée [...] qu'aucune de celles [...] inventées par les hommes," that is as a machine that, being built by God, is much better that any machine made by humans) (Cottingham 1978, 552).

To philosophy, religion must be added. The Greek gods had human features. This fact "reveals their anthropocentrism, the view that human beings are primary and central in the order of things" (Steiner 2005, 1). As Lynn White (1967, 1205) adds (in "the most controversial and widely-discussed paper to approach the environmental crisis from a philosophic-religious perspective") (Sessions 1974, 72): "Christianity is the most anthropocentric religion the world has seen. As early as the 2nd century both Tertullian and Saint Irenaeus of Lyons were insisting that when God shaped Adam, he was foreshadowing the image of the incarnate Christ, the Second Adam. Man [sic] shares, in great measure, God's transcendence of nature. Christianity, in absolute contrast to ancient paganism and Asia's religions (except, perhaps, Zoroastrianism), not only established a dualism of man and nature but also insisted that it is God's will that man exploit nature for his proper ends." From this, White derives the idea that Judeo-Christian thought, maintaining the superiority of human beings over all other forms of life, encouraged the over-exploitation of nature.

According to Robert Livingston Schuyler (1948), anthropocentrism is "Man's Greatest Illusion." Like White, Livingston Schuyler spoke of "man" instead of "human beings" (they were both integrated in the geopolitical discourse of their time, and thus their use of the term "man" to indicate the whole of humanity was "taken-for-granted"). Yet his analysis deserves to be revisited because it adds decidedly profound ideas to the debate. "Illusions of grandeur," he writes, "may take many forms; for example, individu, national, racial, and human" (1948, 48). Precisely on the basis of these illusions, human beings project themselves outside of reality, into that "cartographic reason" (Farinelli 1992) through which they make up another illusion, the illusion of controlling the world (and of being its only master).

Of course, "man" also deludes himself that he is exceptional, above any other living thing. "There are two distinguishable claims implicit in human exceptionalism. The first is that humans are unique, humans are the only beings that do or have X (where X is some activity or capacity); and the second is that humans, by doing or having X, are superior to those that don't do or have X" (Gruen 2011, 4). Point one can be read in many different ways: nonhuman animals cannot write, for example, or have generally more limited forms of communication. However, it is a fact that human beings do not know how to fly, for instance, or return home thanks to their sense of smell. The first point provides grounds for the second, which becomes problematic, as it justifies feeling superior and masterful.

On this basis, a clear distinction between humans and nonhuman animals (dualism) is drawn, whereby people are expected to value humans above other animals. Such an attitude is called "speciesism." The word, coined by Ryder (1970), was later popularized by Peter Singer in his book *Animal Liberation* (1975). Speciesism is used to denote "a prejudice or attitude of bias in favor of the interests of members of one's own species and against those of members of other species." The "anthropocentric discourse" is based on the presumption that human beings are superior to animals in terms of intellectual and linguistic abilities, that they are therefore "exceptional" compared to them, while speciesism affirms the superior right of the species, regardless of the intellectual qualities of individuals. An example offered by Singer (1975) shows how nonhuman animals with greater cognitive abilities than orphaned children with severe brain damage are regularly subjected to invasive and painful research, while orphaned children are not, only because of their biological classification. On the grounds of speciesism, in turn, it is possible to base the right for the exploitation of species of living beings considered inferior.

As will be shown in the following sections, voices against anthropocentrism (and against speciesism) have been raised since antiquity. In recent decades, efforts to overcome anthropocentrism have become increasingly significant. This does not mean that Leopold's "Land Ethic" (1949) has been realized, as Thomas A. Heberlein optimistically wrote in the now distant 1972, underlining how awareness of environmental issues was rapidly increasing at the time. On the contrary, the growth of public concern over environmental problems referred to by Heberlein stems "from a recognition of the consequences such problems can have for human beings rather than the total (nonhuman as well as human) environment" (Dunlap, Van Liere, & Kent 1977, 204). It follows that, even for many of those involved in environmentalism, the growing awareness of environment is linked to the fear that damage to the environment may result in damage to the quality of life of human beings. "To ascribe value to things of nature as they benefit man [sic] is to regard them as instruments to man's [sic] survival or well-being. This is an anthropocentric point of view" (Murdy 1975, 1169). Holding on to this statement, most of the positions expressed today by "mainstream environmentalism," from conservationism to the pursuit of sustainability, can be qualified as anthropocentric. So, despite the great fame achieved by the works of nature writers such as Henry David Thoreau and John Muir, despite the reflections of later authors such as Aldo Leopold, despite the extensive discussion on environmental ethics, deep ecology, ecofeminism, and posthumanism, and despite the growing interest in critical animal studies, anthropocentrism and speciesism remain the dominant attitudes in the 21st century.

2.3 Thinking outside the box

However, it must be stressed that not everyone has always shared anthropocentric positions. Even remaining within the tradition of Western thought, for example, the Homeric writings show a more articulate attitude, suggesting a continuum between humans and nonhuman animals, more than a clear divide (Steiner 2005). This is evidenced by the frequent use of analogies between human warriors and animals of great strength, such as lions; from the various metamorphoses suffered by some of the characters, and by single episodes, such as the exchange of emotion between Argos, the old dog who uses all his strength to wag his tail seeing his master, and Ulysses, who, having arrived home incognito, must hide his emotion, and looks the other way, wiping away "a salty tear."

Some of the pre-Socratic philosophers also advanced non-anthropocentric positions, since they tried "to understand [the human being] as an organism in interaction with the environment, the two being by no means ultimately distinct" (Sessions 1974, 76). In the 6th century BCE, the Pythagoreans did not eat animals, because they believed in metempsychosis, or the transmigration of souls. Around the mid-5th century BCE, Empedocles forbade the sacrifice of animals and meat-eating on a similar basis. In the 1st century CE, Plutarch practiced vegetarianism out of respect for animals; he believed that their experiential capacities were rich enough to prevent us using them for food, and that eating animals was not a natural behavior, but an acquired habit. In the 3rd century, the neo-Platonic Porphyry wrote *On Abstinence from Killing Animals*, where he argues that the practice of eating meat is bad for the health, that no animal sacrifices should be made, and that every being who has perceptions and memory deserves justice (Steiner 2005).

In Lynn White's (1967) opinion, alternative visions, "which might provide an antidote to the 'arrogance' of a mainstream tradition steeped in anthropocentrism" (Brennan and Lo 2020) existed even within Christianity. In the 15th century, for instance, Saint Francis of Assisi, suggested "an alternative Christian view of nature" substituting "the idea of the equality of all creatures, including man [sic], for the idea of man's [sic] limitless rule of creation" (White 1967, 1207).

Outside of Christianity, the Dutch philosopher of Jewish religion Baruch Spinoza (1632–1677) argued against a human-centered view of the universe, implying that "our value systems and judgments are usually the result of seeing things in terms of our own narrow interests stemming from our finite human perspective" (Sessions 1974, 79). Spinoza also does not deny that animals have emotions and that they share with the human being the "endeavor to persist in its own being." This openness does not, however, prevent him from having a perspective marked by speciesism (and also by a strong gender bias), as can be seen from the following statement: "the law against slaughtering animals is based more on empty superstition and effeminate pity than on sound reason." He goes on: "I do not deny that beasts have feelings, but I do deny that it is impermissible, on this account, for us to consult our own advantage, and to use them as we wish and to treat them in such a way as is more convenient for us" (Strawser 2011, 9–10). Later on, David Hume (1711–1776), in his *A Treatise of Human Nature* (1739–40) states that animals are endowed with thought and reason, and pride and humility, and love and hate; but they do not have morality (Arnold 1995).

More advanced in this line of thought are the positions of Jeremy Bentham (1748–1832), starting from an "empiricist" position, who affirms, with a sentence destined to become famous in the world of animal studies and beyond, that: "The French have already discovered that the blackness of the skin is no reason why a human being should be abandoned without redress to the caprice of a tormentor. It may one day come to be recognized that the number of the legs, the villosity of the skin, or the termination of the os sacrum are reasons equally insufficient for abandoning a sensitive being to the same fate [...] the question is not, Can they reason? nor, Can they talk? but, Can they suffer?" (Bentham 1789, Ch. 17, Footnote b).

The German philosopher Arthur Schopenhauer (1788–1860) is "one of the first Western philosophers to accord not only moral standing but moral rights to animals" (Puryear 2017). As Steiner points out (2005, 188), according to Schopenhauer, we owe animals "not mercy but justice." Animals are in all essential aspects identical to us, even if they are less intelligent. For this reason, they cannot continue to appear as creatures without rights. In this respect, Europe has much to learn from the "superior morality of the East, because Brahmanism and Buddhism do not limit their precepts to 'neighbours' but take 'all living beings' under their protection." On the other hand, he is convinced that suffering increases with increasing clarity of conscience, and that therefore the pain suffered by animals is less than that suffered by humans (Puryear 2017). On this basis, he takes a stance against vivisection, unless it is a practice linked to a very important investigation and of immediate use. With regard to the consumption of meat, the death of the animal must be immediate and unexpected. All activities that exploit animals for fun, such as hunting or bullfighting, should be banned (Steiner 2005).

Also, on the other side of the Atlantic, "glimmerings of an alternative viewpoint first appeared, developed by writers who saw that nature could serve as an end in itself and was not merely something to be conquered or exploited" (Payne 1996, 13). One of those writers was Henry David Thoreau (1817–1862), the author of *Resistance to Civil Government* (1849), *Walking* (1851) and *Walden. Or Life in the Woods* (1854). He has been called "an American analogue to Schopenhauer" (Oelschlaeger 1991, 133), because he shares with Schopenhauer a generalized criticism towards Modernism. He also shares with the German philosopher a similar attention to animal lives, even if his interests are mostly focused "in the transcendental implications of nature" (Whitford and Whitford 1951, 293). *Walden* is an autobiographical book, in which the author tells about his own experience of life alone in the woods, taking the opportunity to express a critical attitude towards the consumerism of Western society. At the same time, he exalts the ability to live alone, without needing others, to meditate and spiritually immerse oneself in nature. Because of his scientific approach, Thoreau has been defined a "pioneer ecologist and conservationist" (Whitford and Whitford, 1951) and "the father of American phenology" (Leopold and Jones 1947, 83). Moreover, he "had the brilliance to recognize, before Darwin published his theory of evolution, an organic connection between *Homo sapiens* and nature, a natural world from which the species had come and to which it was bound" (Oelschlaeger 1991, 133): more specifically, his "great genius" was able not only to see the affinity between wild nature and the human being, but also to question the alleged dichotomy between primitive and civilized (Oelschlaeger 1991, 134). *Walden* was highly appreciated by Ralph Waldo Emerson (who actually owned Walden Pond), enjoyed moderate success from the first edition, but was later forgotten (Dean and Scharnhorst 1990). It gradually regained popularity as from the 1890s, and later became a classic of nature-writing (Buell 1995).

Charles R. Darwin (1809–1882), with the theory of descent from a common progenitor (*On the Origin of Species*, 1859), definitively overcomes the idea of a clear separation between species, requiring humans to accept their "animal" nature. In Darwin's perspective, human beings and animals, descending from a common origin, have similar minds, and similar emotions, even though human beings have developed greater intellectual and emotional capacities. Superior animals also possess language skills, again developed to different degrees, from the limited abilities of apes, to those immensely more evolved of humans. Given this conception of "continuity," and given the enormous influence exerted among contemporaries and later, Darwin's theories "provided sufficient evidence to finally inter the idea that nature exists to serve man [sic]" (Murdy 1975, 1168). Consequently, they had the potential to undermine anthropocentrism as an ideology, since they contributed to destroy "humanity's self-image as a purposeful creation at the center of the universe with the exceptional capacity of free will" (Weitzenfeld and Joy 2014, 7).

It did not happen. However, since the publication of *On the Origin of Species*, Darwin had a great influence on the vision of nature of many scholars. One of them was John Muir, perhaps one of the first and most attentive American readers of his works. John Muir can certainly be listed, along with Thoreau, among the American writers who have contributed to providing "glimmerings of an alternative viewpoint" (Payne 1996, 13), but he was also something more: he was the first to "combine esthetic, ecological, economic, and ethical rationales into a persuasive polemic for political change" (Payne 1996, 14).

John Muir (1838–1914) is an iconic figure in the history of environmentalism. Born in Scotland, he emigrated as a child to the US, where, albeit never graduating, he became a respected botanist, geologist, and glaciologist. He was also an explorer, and a writer. He was a co-founder of the Sierra Club in 1892 and one of the first supporters of the idea of the national park in the US (so much so that he is frequently referred to as "the father of American National Parks") (Giblett 2011). Like Thoreau, Muir insisted on the intrinsic worth of nature. In the opinion of both, nature should be preserved for its own sake (Dunlap and Mertig 1991, 209–210). Although he did not leave any properly philosophical treatises, John Muir nevertheless developed theoretical positions in which he constantly challenged the Judeo-Christian derived anthropocentrism and its affirmation within modernist thinking. Specifically, Muir attacks the selfishness of human civilization and calls the human being "Lord Man." Consequently, in the journal he kept of his 1867 walk from Indiana to the Gulf of Mexico (*A Thousand-Mile Walk to the Gulf*, 1916, 122), he writes: "I have precious little sympathy for the selfish propriety of the civilized man, and, if a war of races should occur between the wild beasts and Lord Man, I would be tempted to sympathize with the bears."

In developing a biocentric attitude, Muir realized that "humankind enjoyed no special dispensation, and therefore he abandoned the doctrine of special creation and any supernaturalistic account of the human soul" (Oelschlaeger 1991, 191). In nature, God was incarnate; more precisely, God is nature, "a sacred living temporal presence in *everlasting* process" (Oelschlaeger 1991, 191). Within this process, the human species is no longer the chosen species, just one of many.

2.4 Conservation / preservation

The origins of contemporary environmentalism may be traced back to the conservation movement that emerged in the late 1800s in the United States in response to the exploitation of the country's natural resources.

A "conservation 'ethic' — if not a unified, coherent movement — existed long before. That ethic derived from the fear that abuse of the land threatened the future of American civilization" writes Pisani (1985, 341), and he adds: "Concern for the future of American forests antedated the Civil War. For example, in both *The Pioneers* (1823) and *The Prairie* (1827), James Fenimore Cooper lamented the destruction of trees" (1985, 342). The contributions that have given a more important impulse to the maturation of an environmentalist sensibility, however, are those of George P. Marsh, Henry David Thoreau, and John Muir, all published during the second half of the century.

George Perkins Marsh (1801–1882) was a geographer, diplomat, and politician. His role in the making of the environmental movement in the US was critical to the point that he has been called one its "fathers." In 1847, he gave a speech to the Agricultural Society of Rutland County, Vermont, drawing attention to the destructive impact of deforestation and the consequences that this impact had on the environment. He said that "The inconveniences resulting from a want of foresight in the economy of the forest are already severely felt in many parts of New England, and even in some of the older towns in Vermont. Steep hill-sides and rocky ledges are well suited to the permanent growth of wood, but when in the rage for improvement they are improvidently stripped of this protection, the action of sun and wind and rain soon deprives them of their thin coating of vegetable mold, and this, when exhausted, cannot be restored by ordinary husbandry. They remain therefore barren and unsightly blots, producing neither grain nor grass, and yielding no crop but a harvest of noxious weeds, to infest with their scattered seeds the richer arable grounds below. But this is by no means the only evil resulting from the injudicious destruction of the woods [...]" (Marsh 1848, 18). And then he added that "[I]n many European countries, the economy of the forest is regulated by law; but here, where public opinion determines, or rather in practice constitutes law, we can only appeal to an enlightened self-interest to introduce the reforms, check the abuses, and preserve us from an increase of the evils I have mentioned" (Marsh 1848, 19).

His most important work was published a few years later (1864) with the title *Man and Nature; or, Physical Geography as Modified by Human Action*. Here, he "indicted humankind for wantonly slaughtering animals," and destroying nature's largesse; at the same time, he was certain of "humankind's superiority over the remainder of creation" (Oelschlaeger 1991, 108), and of its capacity to damage it. His positions were thus clearly anthropocentric, but equally clearly capable of highlighting the role of the human being in ruining the natural life of the planet. Today, the "distinction Marsh posited between man [sic] and 'brute' creation is abhorrent to most environmentalists," writes his biographer, David Lowenthal (2000, 17). Still, Marsh demonstrated that humanity is "a destabilizing environmental force whose impacts portended an uncertain future" (Oelschlaeger 1991, 107) and, therefore, that it should also "develop a stewardship of natural resources" (Brulle 1996, 68). Thus, he may be called not only a prophet of conservationism, but also a prophet of the Anthropocene discourse (Lowenthal 2016).

"More than Marsh had dreamed, *Man and Nature* ushered in a revolution in how people conceived their relations with the earth," writes Lowenthal (2000, 3). More specifically, the book "initiated a radical reversal of environmental attitudes. In tandem with the tree-planting crusade that swept the United States in the Arbor Day movement, Marsh's warnings led the American Association for the Advancement of Science in 1873 to petition Congress for a national forestry commission. From this emerged a forest reserve system in 1891, then watershed protection, eventually a federal conservation program for natural resources" (ibid., 4).

Marsh's book was judged "epoch-making" also by Gifford Pinchot (Lowenthal 2009), the man who in fact put Marsh's suggestions about a more prudent management of resources into practice and who is therefore called "the architect of the conservation movement of the early twentieth century" (Miller 2001, 1). A good friend of Theodore Roosevelt, Gifford Pinchot (1865–1946) was the first head of the United States Forest Service, which he founded in 1905. He is usually considered the main promoter of "utilitarian conservationism," because his vision of "managed" conservation "basically meant that lands owned by the federal government could not only be used for recreation by the general public but could also be used, responsibly, by industry for logging, mining and many other purposes including extensive scientific research on tens of thousands of acres of land."[1]

Pinchot's positions on conservation were quite different from those of the other prominent figure of early environmentalism, John Muir. The two men were actually friends for part of their lives, and collaborated for a while (Clayton 2019), but arrived at a clash over the proposed construction of the O'Shaughnessy Dam in the Hetch Hetchy Valley, located in a remote northern section of Yosemite National Park. While Muir opposed the flooding of the valley because of its beauty, Pinchot argued that the beauty of the landscape was less important than its usefulness as a reservoir for San Francisco. The dam was built in 1913 and the friendship split, bringing with it the fracture of the first environmentalist movement, which divided along two lines (Miller 2001).

So, the he story of the conservation movement in the US is usually told by structuring it on the dichotomy between the two "radically contrasting views"[2] of Pinchot and Muir, conservation versus preservation (Schmidt di Friedberg 2004).

"The word conservation first took on its generally understood meaning, the protection, management, and controlled use of natural resources" (Payne 1996). It was connected with "the movement to construct reservoirs to conserve spring flood waters for use later in the dry season," and with "the concept of planned and efficient progress which lay at the heart of the conservation idea" (Hays 1959, 5). According to Philip P. Wells, of the National Conservation Commission, the main aim of "conservation" was "the use of foresight and restraint in the exploitation of the physical sources of wealth as necessary for the perpetuity of civilization, and the welfare of present and future generations" (Hays 1959, 123). By 1909, Pinchot claimed "to have invented the concept of conservationism" (Miller 2001, 153). Concerned by the lack of coordination between the various agencies that dealt in a fragmented way with water, soil, wildlife, forests, etc., on the territory of the United States, he understood that "there was a unity in this complication" and that what could find "order out of chaos was a belief in the power of conservation to transform human activity" (Miller 2001, 154). Specifically, he called conservation "the key to the future," a key whose goal was to produce "the greatest good, for the greatest number, for the longest run." The Pinchot faction of the movement therefore "appropriated the term to mean the comprehensive and well-planned management of natural resources of every character, based on sound ethical and economic grounds" (Payne 1996). While Pinchot appropriated the term "conservation" for his viewpoint, there were, on the other side of the schism, those "who would preserve undeveloped land for its esthetic, spiritual, and recreational values as wilderness" (Nash 1967, 427) and who poured all their energies into the National Parks movement. For this reason, the term "preservation" is usually used to define "the movement to set aside areas of natural scenery or wilderness for appreciation and enjoyment" (Oravec 1981, 245).

Today, the distance between these two components of environmentalism seems to have narrowed: despite much of the literature tending to portray conservationists as "bad," and preservationists as "good," and despite John Muir's sincere biocentrism, both conservation, understood as the careful management of "nature" as a resource, and preservation, understood as respect for "nature" in its wilderness as a spiritual source for human beings, are anthropocentric in their essence and both fall under the umbrella of mainstream environmentalism.

2.5 Challenging anthropocentrism: biocentrism, ecocentrism, deep ecology

In the Western tradition, discordant positions regarding the relationship that the human being has with his/her surroundings have existed for centuries and have manifested themselves even within the most "anthropocentric" of religions; however, a philosophical discussion on these issues, aimed at achieving a "basic change of values," has developed under the definition of "environmental ethics" in relatively recent times in the Anglophone world.

The so-called "father of environmental ethics" is Aldo Leopold, who, in the essay "The Land Ethic," wrote: "There is as yet no ethic dealing with man's [sic] relation to land and to the animals and plants which grow upon it [...] The extension of ethics to this third element in human environment is [...] an evolutionary possibility and an ecological necessity" (Leopold 1949, 238–239). However, the field acquired a commonly agreed upon "name" only 30 years later, in 1979, with the foundation of the journal *Environmental Ethics* (Hargrove 1992).

"Environmental Ethics is concerned with the values attached to the natural world. It deliberates on the appropriate ethical stance to be adopted by humans in order to protect or promote these values" (Tharakan, Iype, and Afonso 2011,

27). Its articulated discussion is very rich and nuanced, but it must be remarked that "The opposition between anthropocentric and ecocentric or biocentric approaches" still captures much of the debate (Kopnina 2012, 11).

The tenets of biocentrism can be summarized as follows: "each animal and plant in the natural world pursues its own good in its own way and therefore is similar, in that respect, to a human" (Taylor 1983, 237); "the Earth's biosphere is one total system of interdependent parts, the good of each part being dependent on the integrity of the whole" (ibid., 238); and, eventually, "humans have no privileged place in this whole ecological scheme of things" (ibid.).

"Biocentrism" is sometimes used synonymously with ecocentrism. Still, there are significant differences between the two approaches, about "the focus on the 'unit' of study or care — be it individual species, individuals within the species, or entire habitats with their biota" (Kopnina, Washington, Taylor, and Piccolo 2018, 114). Specifically, ecocentrism also includes in its realm nonliving elements of the environments, with an "acknowledgment of value in geology and geomorphology of the land itself."

In consequence, John Muir, the champion of a non-anthropocentric vision of nature, who believed that a human being is only one of the components in a non-hierarchical natural world, in which there is no distinction in importance between living species, was certainly one of the first promoters of a biocentric approach. The "trailblazer" of ecocentrism, rather, is Aldo Leopold, who wrote that: "the land ethic simply enlarges the boundaries of the community to include soils, waters, plants, and animals, or collectively: the land" (Leopold 1949, 204). Leopold's fundamental point is that humans should remove the distinction between people and nature and recognize that they are simply members of the ecological community. On this basis, together with Thoreau and Muir, he is considered "the third giant of wilderness philosophy" (Oelschlaeger 1991, 205).

He differs from them, however, because he attributes to human beings, whom he defines as an element of nature, the duty to preserve the integrity, stability, and beauty of natural systems. And he differs from utilitarian conservationists, because he believes that efficiency and utility are important but secondary aspects. He also differs profoundly from many supporters of contemporary animal rights because he considers "game management" to be fundamental in the ecological field, and "game management implies the associated activities of hunting and killing game animals for human recreation" (Oelschlaeger 1991, 218).

Still today, pro-hunting rhetoric continues to use as arguments in its favor the sustainability of the selection of game species and the protection of the habitat from the risk of excessive numerical growth of a species not preyed upon. The elements brought by Aldo Leopold to defend hunting were based on these very points. Moreover, "according to those who knew him, Leopold saw hunting as an expression of love for the natural world and, at the very least, an excellent way to develop a personal connection with the land" (Simpson and Cain 2000, 183). Of course, anti-hunters "do not judge the correctness of hunting by the hectares of protected game habit, the receipt totals of tourism revenues, or even the health of animal population" (ibid., 182). They judge the practice only "by the pain inflicted upon individual animals and the morality of any human sport that causes such pain" (ibid.). From these different positions arises the "animal rights/environmental ethics debate" (see Hargrove 1992), which focuses on the question of whether it is possible for an animal welfare ethic to also provide an appropriate foundation for an environmental ethic pertaining to wild animals, and not only to domestic animals exposed to factory farming or experimentation.

Biocentrism and ecocentrism are "the ultimate norm" of deep ecology (Devall 1991, 248), an environmental movement that wants to place itself outside of mainstream environmentalism. The expression "deep ecology" is connected with the work of Arne Naess, a Norwegian activist, who, in a 1973 article (Naess 1973, 97), defined as "shallow ecology" the "fight against pollution and resource depletion," whose "central objective" is "the health and affluence of people in the developed countries." On the other hand, "deep ecology" is a normative, eco-philosophical movement based on the idea of "ecocentric identification." According to Devall, "Humans are one of myriad self-realizing beings, and human maturity and self-realization come from broader and wider self-identification. Out of identification with forests, rivers, deserts, or mountains comes a kind of solidarity: 'I am the rainforest' or 'I am speaking for this mountain because it is a part of me'." (Devall 1991, 248).

In the definition offered by the *Encyclopedia of Environmental Ethics and Philosophy*, "deep ecology [...] refers to an egalitarian and holistic environmental philosophy founded on phenomenological methodology. By way of direct experience of nonhuman nature, one recognizes the equal intrinsic worth of all biota as well as one's own ecological interconnectedness with the lifeworld in all its plenitude" (Keller 2009, 206).

The basic principle of deep ecology is "the equal right to live and blossom" for all living entities. It assumes that all forms of life have a non-gradual inherent value, and includes in the idea of life rivers, mountains, and landscapes. Because of this broader definition of "living beings," deep ecology refers preferably to the concept of "ecosphere" instead of "biosphere."

The positions of Naess were later enriched by the contribution of other thinkers, such as Devall (1980, 1991), and Sessions (Devall and Sessions 1985), and achieved a widespread popularity, which spanned "from headline grabbing environmental activists dressed in coyote costumes to scholars of an astonishing assortment of backgrounds and interests" (Keller 2009, 206). Moreover, if the opposition to anthropocentrism is shared by all those who refer to this approach, the alternatives to anthropocentrism vary, "depending on the non-domination-of-nature tradition(s) to which particular deep ecologists turn" (Sessions 1991, 92), ranging from the thought of the philosopher Spinoza, to that of Heidegger and Buddhism. So, deep ecology, more than a single theoretical position, is an "umbrella concept" used in different ways, but always connected with the desire of opposing anthropocentrism with a biocentric egalitarianism vision.

That has not prevented it from being subjected to major criticism, such as Richard Sylvan (*A Critique of Deep Ecology*, 1985), which disputes the possibility of applying egalitarian biocentrism in a proper sense, applying the same rights in an undifferentiated manner to all species, including viruses and bacteria, or to the physical components of the planet, such as rivers, mountains, waterfalls, sunsets, and so on. Deep ecology was also accused of eco-fascism. Specifically, "Deep ecologists see this vague and undifferentiated humanity essentially as an ugly 'anthropocentric' thing — presumably a malignant product of natural evolution — that is 'overpopulating' the planet, 'devouring' its resources, and destroying its wildlife and the biosphere — as though some vague domain of 'nature' stands opposed to a constellation of nonnatural human beings, with their technology, minds, society, etc. Deep ecology [has been] formulated largely by privileged male white academics" (Bookchin 1987).[3]

So, deep ecology has been heavily criticized also by ecofeminists, because it tackles anthropocentrism, but seems unable to overcome all other "-centrisms," including those of gender.

2.6 Ecofeminism and posthumanism

A step forward in the discussion of the relationship between human beings and "nature" is represented by the affirmation of ecofeminist thinking and the posthumanist approach. From poststructuralism, in general, came first of all the stimulus to deconstruct the concepts of wilderness and nature (the most complex word in the English language, following Raymond Williams' opinion) (1983, 219) and the practice of nature writing.

Many authors suggest the "constructedness" of the idea of "wilderness" (Nash 1982; Callicott and Nelson 1998) as well as that of "nature," often pointing out "the implicit race, class, and gender connotations" (DeLuca and Demo 2001, 543) of the two concepts. As Giblett (2011, 27) underlines, nature writing is connected with natural history, and both are (or have been) "an outdoor school to which only men, especially the 'coming man' of settler modernity, were admitted." The concept of wilderness, in Giblett's analysis, is not something existing *per se*, "a pristine natural place largely free from human or industrial modification," but "a construct of settler societies that project a premodern past onto the hypermodern present."

"Wilderness [...] like nature, is a polysemic sign in search of a referent and [...] entails a disjunction between subject and object" (Giblett 2011, 97). Wilderness is what is not. "Remoteness from access and settlement defines wilderness largely in relation to what it is not rather than in terms of what it is. Similar difficulties afflict definitions of wilderness in terms of absence of human inhabitation, though historically for settler societies this has not troubled the consciences of many" (Giblett 2011, 103).

Moreover, "the interest in wilderness lies somewhere between a western conservationist or tourist and land unmodified by modern industrial technology." The concept of wilderness, therefore, is strongly linked to a certain type of culture (the Western culture of settlers), and consequently of gender and race. About that, one can only agree with the statement that "The Achilles heel of the environmental movement is its whiteness" (DeLuca and Demo 2001, 541). Indeed, "With its focus on wilderness, the traditional environmental movement [...] pretends there were no indigenous people in the North American plains and forests" (ibid., 542).

Also "the concept/referent of nature has a gender politics. Nature has been feminized, and culture masculinized, in Western and other cultures. Culture has been construed and troped in masculine terms and nature in feminine ones in the Western patriarchal tradition" (Giblett 2011, 29).

The deconstruction of the concepts of nature and wilderness makes evident the tendency to create oppositions and dualisms (in terms of gender, race, spatial articulation, etc.) typical of Western thought. So, particularly active in the work of deconstructing the concept of nature have been scholars in the field of cultural studies and feminism, and specifically Carolyn Merchant (1980) and Donna Haraway (1991). Feminists, in particular, are responsible for highlighting the parallelism between the domination of nature and the domination of women as resources (Merchant 1980), and also of the responsibility of the anthropomorphic and stereotypic labels that contribute to the perpetuation of a system repressive to both women and nature. Environmental movements and movements for the liberation of women have some intentions in common, so juxtaposing their goals "can suggest new values and social structures, based [...] on the full expression of both male and female talent and on the maintenance of environmental integrity" (Merchant 1980, 51).

The environmentalist philosophy that has developed within the feminist milieu, however, takes positions stronger than deep ecology. While sharing with it "the desire to supplant the predominant Western anthropocentric environmental frameworks" (Sessions 1991, 90), ecofeminists accuse the ecological movement in general of failing "to make the conceptual connections between the oppression of women and the oppression of nature (and to link these with other systems of oppression)" (Warren 1987, 8). So, "ecofeminism insists that a proper analysis must also emphasize the intimate logical and historical connections between the various forms of domination — the same logic and attitudes of superiority and practices of domination humans (men?) display in their relations toward the nonhuman dimensions of the world are found in men's relations to women and in imperialistic, racist, and classist structures and practices" (Sessions 1991, 95).

Specifically, injustices against human and nonhuman animals cannot be addressed independently, because "they are rooted in hegemonic centrisms — widespread and often unquestioned cultural practices of understanding and evaluating the world through the experiences and norms of an exclusive, elite population" (Plumwood 2002, quoted by Weitzenfeld and Joy 2014, 9). In this perspective, Cudworth (2005) coins the concept of "anthroparchy," which identifies the system of power connecting anthropocentrism, patriarchate, animal exploitations, and capitalist domination of nature.

Ecofeminist positions, such as those developed by D'Eaubonne (1974), and later by Cudworth (2005) and Sturgeon (1997, 2009), questioning constructions of class, race, gender, and nature and all the categories separating "men and women, white people and people of color, humans and animals, mind and body, rationality and emotion, straight people and queer people" (Sturgeon 2009, 9), partly coincide with posthumanism, which "does not employ any frontal dualism or antithesis, demystifying any ontological polarization through the postmodern practice of deconstruction" (Ferrando 2013, 29).

At the root of these mechanisms of separation, differentiation, and dominance, there are the "Western ideological frameworks [which] operate dualistically" (Sturgeon 2009, 9). These borders not only differentiate, but also create a hierarchy, defining who is hierarchically superior and who is below. So, ecofeminism and posthumanism also share the desire to defy a system of "unequal power relations which are part of a social structure which is also capitalist and patriarchal" (Carter and Charles 2011, 21). Instead, an environmental activist positioned on ecocentrism can say: "Racism and Sexism for example are social issues but they are not issues relevant to the survival of the biosphere [...] I think that speciesism is a far more serious issue" (quoted by Kopnina 2012, 13).

Albeit with different emphasis on "who needs justice" (only less powerful people, nonhumans, everyone?) (Kopnina 2018, 202), the efforts of all the scholars who relate to these approaches are aimed at demonstrating how unsustainable are both the idea of human exceptionality and the human-animal dichotomy. Indeed, the line that one wants to draw around the human being is a "mobile border" (Agamben 2002), which "passes first of all within man [sic]" (Agamben 2002, 2004, 16); that is, it is a line that risks becoming an internal divide within humankind itself, which cuts off some of its elements, and produces "an ethico-politically privileged inside and sacrificial outside" (Weitzenfeld and Joy 2014, 8).

2.7 Spatializing ecofeminism / Posthumanizing geo-graphy

The same mechanisms of dualism and hierarchy separating "men and women, white people and people of color, humans and animals, etc." (Sturgeon 2009, 9) may act at the spatial level. Levels of environmental degradation, pollution and other environmental risks may be unequally distributed by race and class (Mohai, Pellow and Roberts 2009). Hence, the idea of the "production for space" (Lefebvre 1974) becomes fundamental to approaching the concepts of "environmental racism" (Bullard 1990) and "environmental justice". Specifically, "environmental justice" is set as the principle that "all people and communities are entitled to equal protection of environmental and public health laws and regulations" (Mohai, Pellow, and Roberts 2009).

The production of space does not act only in a material sense (Teelucksingh 2002). These dualisms, and these mechanism of "naturing" people, may also be used to interpret spatial Otherness in form of "imaginative geographies," as suggested by Edward Said in *Orientalism* (1978). So, contexts too are "natured," and explained in essentializing terms, even though this "idealized" vision of "places [...] (supposedly) inhabited by coherent and homogeneous communities" (Massey 1994, 146), albeit deeply rooted in the Western imagination, does not correspond to the reality.

Indeed, it is typical of the Western geopolitical imagination to spatialize difference, imposing a "binary division of the world" where local differences are "invariably assimilated into a global geographical taxonomy with its roots in Europe" (Agnew 2003, 23). This has led to a "crystallization" of the idea of "The West," in a binary juxtaposition to the "less developed Rest" (Hall 1992), producing a global imagination where "The West and the Rest are two sides of a single coin" (Hall 1992, 278).

This process of crystallization follows four different discursive strategies: "idealization; the projection of fantasies of desire and degradation; the failure to recognize and respect difference; the tendency to impose European categories and norms, to see difference through the modes of perception and representation of the West" (Hall 1992, 308). The ensuing process of stereotyping tends to collapse the characteristics of the two sides, "splitting" the stereotype "into two halves — its 'good' and 'bad' sides; this is [...] dualism" (ibid.). It seems superfluous to add that "the West and the Rest" are "historical and linguistic constructs" (and so their "meanings change over time") (ibid., 279). They are also visually spatialized through the "cartographic reason" (Pickles 2004).

On the basis of the same "geographical imagination," and of the same discursive mechanisms, geographical dualisms differentiate between the world of modernity/culture and the world of premodernity/nature, sometimes exalting the presumed purity of the latter in opposition to the former, in other cases dreaming that from the polluted, degraded, and dirty former, one can escape into the latter (which acts as Planet B). Thus, spatial dualisms must be included in the list of dualisms to be rethought within the ecofeminist / posthumanist theoretical umbrella. Indeed, when Rosi Braidotti says that "'Man' cannot claim to represent all humanity because that 'Man' is a culture-specific, gender-specific, race-specific and class-specific entity: it is a European, male, white, intellectual ideal" (Braidotti and Veronese 2016, 339), she already says something very "geopolitical," albeit perhaps in a manner that is not theoretically explicit. At the same time, geography and geopolitics, beyond questioning constructions of class, race, gender, and nature, should further articulate their posthuman dimension and question the human-nonhuman animal divide in its spatial hues. Positive stimuli in this direction have not been lacking, starting from those offered by Kropotkin (1902), who considered evolution a process of "mutual aid" rather than competition, investigating the role of mutually beneficial cooperation and reciprocity among human and nonhuman species. To this effect are also Élisée Reclus' "cosmological sense of nature" (Guest 2017, 88) and his writings in favor of vegetarianism (1901) (White 2015). Working on the track laid out by Reclus, a small group of anarchist geographers are now moving in this direction, offering a notable exception to the anthropocentric stance of the discipline; among them, Richard White (2015), Simon Springer (White, Springer, and Souza 2016), and the members of the Vegan Geographies Collective (Ferretti 2019).

Inspirational in this direction can also be considered the work of Yi-Fu Tuan (*Dominance and Affection: The Making of Pets*, 1984), where Tuan traces the inherently unequal and patriarchal power relations between human beings and pets, stressing how the abuse of power also manifests itself in the treatment of human members such as children, women, servants, who, at different times, may be both highly valued and severely controlled and trained as mechanical toys.

Moreover, Raffestin (1983, 2012) has repeatedly addressed the different perception of space by human and nonhuman animals and the biophilosophy of Jakob von Uexküll. Indeed, as Raffestin writes "space, as a certain conception of geography would have it, is a mental construction of the human species that maintains relations both with what we can call exteriority and with alterity for the satisfaction of its needs. Other animal species, of course, have their own constructions, as developed by Jakob von Uexküll (1934). Nonhuman animal species also have a representation of their surroundings, without which they would not be able to satisfy their needs. They probably also have a representation of their 'geography', but in contradistinction to that of humans we cannot access this except through experiences and programs that would allow us, in whatever form, to make them explicit (see von Uexküll, 1934)."

However, "it is a peculiar fact that a discipline which, in part, defines itself as the study of society-environment relations has conspicuously failed to engage with questions of the political status of the non-human" (Castree 2003, 207). The study of nonhuman animals and their spatial relations with humans is restricted to a delimited, albeit growing, area of the discipline (animal geographies: see Philo 1995; Wolch and Emel 1995; Wolch and Emel 1998; Philo and Wilbert 2000; Buller 2014; Buller 2015; Buller 2016; Wilcox and Rutherford 2018), while the attempt to develop a "hybrid" disciplinary approach, from the point of view of a more-than-human geography, still remains linked to the commitment of a small number of scholars (Whatmore 2002; Braun 2005; Whatmore 2006; Shingne 2020).

Political geography substantially forgets that "animals are already subjects of, and subject to, political practices," while political ecology tends to assume "nature" as a set of static resources (Hobson 2007). Geopolitics — which throughout the 20th century oscillated from the materialistic approach typical of classical geopolitics to the approach focused on the analysis of representations typical of critical geopolitics (Kelly 2006) — has remained, in both cases, traditionally anthropocentric, even when opening up its interests to environmental security issues.

As Dalby writes (2014, 3) "One of the key dichotomies that structures modern thinking, the division between human and nature, is no longer tenable" because "we are part of a nature that the affluent urbanized fossil fueled part of humanity is rapidly changing." In this perspective, he suggests rethinking geopolitics "to facilitate shifting analysis from focusing on questions of dominance on a divided world to modes of sharing a crowded planet which is actively being transformed by human action" (2014, 15). His idea of sharing, however, remains linked to the fact that sharing involves humanity alone, while posthumanism states that the biosphere/ecosphere is "a system co-constitutive," not only with other humans, but also with nonhuman systems (Cudworth & Hobden 2017). In International Relations, IR, the idea that "the co-evolution of human communities, non-human animals and the 'natural environment' can be understood as interpolated through institutions and practices of biopower that give rise to patterns of multiple complex inequalities" has promoted the making of a "Posthuman International Relations" approach (Cudworth & Hobden 2011). An invitation to a "posthuman turn in geopolitics, incorporating animals, 'nature', and other objects into our understandings of the geopolitical" was also issued by Jason Dittmer in 2014; however, while in IR the debate is lively (Eroukhmanoff and Harker 2017), in geopolitics the challenge has yet to be met.

Notes

1 https://www.fs.usda.gov/features/conservation-versus-preservation
2 https://www.fs.usda.gov/features/conservation-versus-preservation
3 https://libcom.org/library/social-versus-deep-ecology-bookchin/

Bibliography

Agamben, Giorgio (2002). *L'aperto. L'uomo e l'animale.* Bollati Boringhieri (2004. *The Open. Man and Animal.* Stanford University Press).

Agnew, John A. (2003). *Geopolitics: Re-Visioning World Politics.* Routledge.

Arnold, Denis G. (1995). Hume on the Moral Difference between Humans and Other Animals. *History of Philosophy Quarterly* 12(3), 303–316. www.jstor.org/stable/27744668

Bentham, Jeremy (1780–89). *An Introduction to the Principles of Morals and Legislation.* Clarendon Press.

Bookchin, Murray (1987). Social Ecology Versus Deep Ecology: A Challenge for the Ecology Movement. *Green Perspectives: Newsletter of the Green Program Project* 4–5.

Braidotti, Rosi, & Veronese, Cosetta (2016). Can the Humanities Become Posthuman? A Conversation. In Opperman, Serpil, & Iovino, Serenella (eds.), *Environmental Humanities. Voices from the Anthropocene* (pp. 339–346). Rowman & Littlefield.

Braun, Bruce (2005). Environmental Issues: Writing a More-Than-Human Urban Geography. *Progress in Human Geography* 29(5), 635–650. https://doi.org/10.1191/0309132505ph574pr

Brennan, Andrew, & Lo, Yeuk-Sze (2020). Environmental Ethics. In Zalta, Edward (ed.), *The Stanford Encyclopedia of Philosophy.* https://plato.stanford.edu/archives/sum2020/entries/ethics-environmental/.

Brulle, Robert J. (1996). Environmental Discourse and Social Movement Organizations: A Historical and Rhetorical Perspective on the Development of US Environmental Organizations. *Sociological Inquiry* 66(1), 58–83. https://doi.org/10.1111/j.1475-682X.1996.tb00209.x

Buell, Lawrence (1995). *The Environmental Imagination: Thoreau, Nature Writing, and the Formation of American Culture*. Harvard University Press.

Bullard, Robert (1990). *Dumping in Dixie: Race, Class, and Environmental Quality*. Westview.

Buller, Henry (2014). Animal Geographies I. *Progress in Human Geography* 38(2), 308–318. https://doi.org/10.1177/0309132513479295

Buller, Henry (2015). Animal Geographies II: Methods. *Progress in Human Geography* 39(3), 374–384. https://doi.org/10.1177/0309132514527401

Buller, Henry (2016). Animal Geographies III: Ethics. *Progress in Human Geography* 40(3), 422–30. https://doi.org/10.1177/0309132515580489

Callicott, J.Baird, & Nelson, Michael P. (eds.) (1998). *The Great New Wilderness Debate*. University of Georgia Press.

Carter, Bob, & Charles, Nickie (eds.) (2011). *Human and Other Animals: Critical Perspectives*. Springer.

Castree, Noel. (2003). Environmental Issues: Relational Ontologies and Hybrid Politics. *Progress in Human Geography* 27(2), 203–211. https://doi.org/10.1191/0309132503ph422pr

Clayton, John (2019). *Natural Rivals*. Simon and Schuster.

Cottingham, John (1978). 'A Brute to the Brutes?': Descartes' Treatment of Animals. *Philosophy* 53(206), 551–559.

Cudworth, Erika (2005). *Developing Ecofeminist Theory: The Complexity of Difference*. Springer.

Cudworth, Erika, & Hobden, Stephen (2011). *Posthuman International Relations: Complexity, Ecologism and Global Politics*. Zed Books.

Cudworth, Erika, & Hobden, Stephen (2017). Post-Human Security. In Burke, Anthony, & Parker, Rita (eds.), *Global Insecurity. Futures of Global Chaos and Governance* (pp. 65–81). Palgrave Macmillan.

Dalby, Simon (2014). Environmental Geopolitics in the Twenty-First Century. *Alternatives: Global, Local, Political* 39(1), 3–16. https://doi.org/10.1177/0304375414558355

d'Eaubonne, Francoise (1974). *Le féminisme ou la mort*. FeniXX.

Dean, Bradley P., & Scharnhorst, Gary (1990). The Contemporary Reception of 'Walden.' *Studies in the American Renaissance*, 293–328. www.jstor.com/stable/30227595

DeLuca, Kevin, & Demo, Anne (2001). Imagining Nature and Erasing Class and Race: Carleton Watkins, John Muir, and the Construction of Wilderness. *Environmental History* 6(4), 541–560.

Devall, Bill (1980). The Deep Ecology Movement. *Natural Resources Journal* 20(2), 299–322.

Devall, Bill (1991). Deep Ecology and Radical Environmentalism. *Society & Natural Resources* 4(3), 247–258. https://doi.org/10.1080/08941929109380758

Devall, Bill, & Sessions, George S. (1985). *Deep Ecology: Living as if Nature Mattered*. Peregrine Smith.

Dittmer, Jason (2014). Geopolitical Assemblages and Complexity. *Progress in Human Geography* 38(3), 385–401. https://doi.org/10.1177/0309132513501405

Dunlap, Riley E., & Val Liere, Kent D. (1977). Land Ethic or Golden Rule: Comment on 'Land Ethic Realized' by Thomas A. Heberlein, JSI, 28(4), 1972. *Journal of Social Issues* 33(3), 200–207. https://doi.org/10.1111/j.1540–4560.1977.tb01891.x

Dunlap, Riley E., & Mertig, Angela G (1991). The Evolution of the US Environmental Movement from 1970 to 1990: An Overview. *Society & Natural Resources* 4(3), 209–218.

Eroukhmanoff, Clara, & Harker, Matt (2017). *Reflections on the Posthuman in International Relations*. E-International Relations.

Farinelli, Franco (1992). *I segni del mondo. Immagine cartografica e discorso geografico in età moderna*. Academia Universa Press.

Ferrando, Francesca (2013). Posthumanism, Transhumanism, Antihumanism, Metahumanism, and New Materialisms. *Existenz* 8(2), 26–32. existenz.us › Vol.8-2Ferrando.pdf

Ferretti, Federico (2019). Anarchism/Anarchist Geographies. In Kobayashi, Audrey (ed.), *International Encyclopaedia of Human Geography*, 2nd ed. (pp. 119–126). Elsevier.

Giblett, Rod (2011). *People and Places of Nature and Culture*. The University of Chicago Press.

Gruen, Lori (2011). *Ethics and Animals: An Introduction*. Cambridge University Press.

Guest, Bertrand (2017). Environmental Awareness and Geography: Reading Reclus Ecocritically? In Specq, François (ed.), *Environmental Awareness and the Design of Literature* (pp. 69–89). Brill Rodopi.

Hall, Stuart (1992). The West and the Rest: Discourse and Power. In Hall, Stuart, & Gieben, Bran (eds.), *Formations of Modernity* (pp. 275–331). Polity Press.

Haraway, Donna (1991). *Simians, Cyborgs, and Women: The Reinvention of Nature*. Routledge.

Hargrove, Eugene C. (ed.) (1992). *The Animal Rights/Environmental Ethics Debate. The Environmental Perspective*. State University of New York Press.

Hays, Samuel P. (1959). *Conservation and the Gospel of Efficiency: The Progressive Conservation Movement, 1890–1920*. Harvard University Press.

Hayward, Tim (1997). Anthropocentrism: a Misunderstood Problem. *Environ. Value* 6(1), 49–63. https://doi.org/10.3197/096327197776679185.

Heberlein, Thomas A. (1972). The Land Ethic Realized: Some Social Psychological Explanations for Changing Environmental Attitudes. *Journal of Social Issues* 28(4), 79–87. https://doi.org/10.1111/j.1540-4560.1972.tb00047.x

Hobson, Kersty (2007). Political Animals? On Animals as Subjects in an Enlarged Political Geography. *Political Geography* 26(3), 250–267. [https://doi.org/10.1016/j.polgeo.2006.10.010

Iovino, Serenella (2012). Material Ecocriticism: Matter, Text, and Posthuman Ethics. In Müller, Timo, & Sauter, Michael (eds.), *Literature, Ecology,* Ethics*: recent trends in ecocriticism* (pp. 51–68). Winter Verlag.

Keller (2009). Deep Ecology. In Callicott, J. Baird, & Frodeman, Robert. *Encyclopedia of Environmental Ethics and Philosophy* (vol. 1, pp. 206–211). Macmillan Reference USA.

Kelly, Phil (2006). A Critique of Critical Geopolitics. *Geopolitics* 11(1), 24–53. [https://doi.org/10.1080/14650040500524053]

Kopnina, Helen N. (2012). Re-Examining Culture/Conservation Conflict: the View of Anthropology of Conservation through the Lens of Environmental Ethics. *Journal of Integrative Environmental Sciences* 9(1), 9–25. https://doi.org/10.1080/1943815X.2011.625951

Kopnina, Helen N. (2018). Just Conservation: In Defense of Environmentalism. In Dhiman, Satinder, & Marques, Joan (eds.), *Handbook of Engaged Sustainability: Contemporary Trends and Future Prospects* (pp. 201–220). Springer.

Kopnina, Helen, Washington, Haydn, Taylor, Bron, & Piccolo, John J. (2018). Anthropocentrism: More than just a Misunderstood Problem. *Journal of Agricultural and Environmental Ethics* 31(1), 109–127. https://doi.org/10.1007/s10806-018-9711-1

Kropotkin, Petr A. (1902). *Mutual Aid: A Factor of Evolution*. McClure, Philips & Company.

Lefebvre, Henri (1974). *La production de l'espace*. Anthropos (1991, *The Production of Space*. Monoskop).

Leopold, Aldo (1949). *A Sand County Almanac and Sketches Here and There*. Oxford University Press.

Leopold, Aldo, & Jones, Sara Elizabeth (1947). A Phenological Record for Sauk and Dane Counties, Wisconsin. *Ecological Monographs* XVII, 81–122. JSTOR, www.jstor.org/stable/1948614

Livingston Schuyler, Robert (1948). Man's Greatest Illusion. *Proceedings of the American Philosophical Society* 92(1), 46–50. http://www.jstor.com/stable/3143626

Lowenthal, David (2000). Nature and Morality from George Perkins Marsh to the Millennium. *Journal of Historical Geography* 26(1), 3–27. https://doi.org/10.1006/jhge.1999.0188

Lowenthal, David (2009). *George Perkins Marsh: Prophet of Conservation*. University of Washington Press.

Lowenthal, David (2016). Origins of Anthropocene Awareness. *The Anthropocene Review* 3(1), 52–63. https://doi.org/10.1177/2053019615609953

Marafiote, Tracy, & Plec, Emily (2011). From Dualisms to Dialogism: Hybridity in Discourse about the Natural World. In Depoe, Stephen P. (ed.), *The Environmental Communication Yearbook* (3, pp. 49–75). Routledge.

Marsh, George Perkins (1848). *Address delivered before the Agricultural society of Rutland County, Sept. 30, 1847*. The Herald Office.

Marsh, George Perkins (1864). *Man and Nature; or, Physical Geography as Modified by Human Action*. Charles Scribner.

Massey, Doreen (1994). *Space, Place and Gender*. Polity Press.

Merchant, Carolyn (1980). *The Death of Nature: Women, Ecology, and the Scientific Revolution*. Harper and Row.

Massey, Doreen (1994). *Space, Place and Gender*. Polity Press.

Miller, Char (2001). *Gifford Pinchot and the Making of Modern Environmentalism*. Island Press.

Milstein, Tema (2009) "Somethin' Tells Me It's All Happening at the Zoo": Discourse, Power, and Conservationism. *Environmental Communication* 3(1), 25–48 https://doi.org/10.1080/17524030802674174

Mohai, Paul, Pellow, David, & Roberts, J. Timmons (2009). Environmental Justice. *Annual Review of Environment and Resources* 34, 405–430. https://doi.org/10.1146/annurev-environ-082508-094348

Muir, John (1916). *A Thousand-Mile Walk to the Gulf* (Edited by William Fredric Bade). Houghton Mifflin Company.

Murdy, William H. (1975). Anthropocentrism: A Modern Version. *Science* 187(4182), 1168–1172. [https://doi.org/0.1126/science.187.4182.1168]

Naess, Arne (1973). The Shallow and the Deep, Long-Range Ecology Movement. A Summary. *Inquiry. An Interdisciplinary Journal of Philosophy and the Social Sciences* 16, 95–100.

Nash, Roderick (1967). John Muir, William Kent, and the Conservation Schism. *Pacific Historical Review* 36(4), 423–433. https://online.ucpress.edu › phr › article-abstract › 36 › 4

Nash, Roderick (1982). *Wilderness and the American Mind*. Yale University Press.

Oelschlaeger, Max (1991). *The Idea of Wilderness: From Prehistory to the Age of Ecology*. Yale University Press.

Oravec, Christine (1981). John Muir, Yosemite, and the Sublime Response: A Study in the Rhetoric of Preservationism, *Quarterly Journal of Speech* 67(3), 245–258. https://doi.org/10.1080/00335638109383570

Payne, Daniel G. (1996). *Voices in the Wilderness: American Nature Writing and Environmental Politics*. University Press of New England.

Philo, Chris (1995). Animals, Geography and the City: notes on Inclusions and Exclusions. *Environment and Planning D: Society and Space* 13(6), 655–681. https://doi.org/10.1068/d130655

Philo, Chris, & Wilbert, Chris (eds.) (2000). *Animal Spaces, Beastly Places. New Geographies of Human-Animal Relations*. Routledge.

Pickles, John (2004). *A History of Spaces: Cartographic Reason, Mapping, and the Geo-Coded World*. Psychology Press.

Pisani, Donald J. (1985). Forests and Conservation, 1865–1890. *The Journal of American History* 72(2), 340–359. https://doi.org/10.2307/1903379

Plumwood, Val (2002). *Feminism and the Mastery of Nature*. Routledge.

Puryear, Stephen (2017). Schopenhauer on the Rights of Animals. *European Journal of Philosophy* 25(2), 250–269. https://doi.org/10.1111/ejop.12237

Raffestin, Claude (ed.) (1983). *Geografia Politica: teorie per un progetto sociale*. Unicopli.

Raffestin, Claude (2012). Space, Territory, and Territoriality. *Environment and Planning D: Society and Space* 30(1), 121–141. https://doi.org/10.1068/d21311

Ryder, Richard D. (1970). Speciesism. Leaflet.

Said, Edward (1978). *Orientalism*. Pantheon Books.

Schmidt di Friedberg, Marcella (2004). *L'arca di Noè. Conservazionismo tra natura e cultura*. Giappichelli.

Sessions, George S. (1974). Anthropocentrism and the Environmental Crisis. *Humboldt Journal of Social Relations* 2(1), 71–81. www.jstor.org/stable/23261527

Sessions, Robert (1991). Deep Ecology versus Ecofeminism: Healthy Differences or Incompatible Philosophies?. *Hypatia* 6(1), 90–107. https://doi.org/10.1111/j.1527-2001.1991.tb00211.x

Shingne, Marie Carmen (2020). The More-than-Human Right to the City: A Multispecies Reevaluation. *Journal of Urban Affairs*. [https://doi.org/10.1080/07352166.2020.1734014]

Simpson, Steven V., & Cain, Kelly D. (2000). Recreation's Role in the Environmental Ethics Dialogue: the Case of Aldo Leopold and the Morality of Hunting. *Leisure/Loisir* 25(3–4), 181–197. https://doi.org/10.1080/14927713.2000.9649916

Singer, Peter (1975). *Animal Liberation: A New Ethics for our Treatment of Animals*. HarperCollins.

Steiner, Gary (2005). *Anthropocentrism and Its Discontents: The Moral Status of Animals in the History of Western Philosophy*. University of Pittsburgh Press.

Strawser, Michael (2011). *On the Specter of Speciesism in Spinoza*. North American Spinoza Society.

Sturgeon, Noël (1997). *Ecofeminist Natures: Race, Gender, Feminist Theory and Political Action*. Routledge.

Sturgeon, Noël (2009). *Environmentalism in Popular Culture: Gender, Race, Sexuality, and the Politics of the Natural*. University of Arizona Press.

Sylvan, Richard (1985). *A Critique of Deep Ecology*. Australian National University.

Taylor, Paul W. (1983). In Defense of Biocentrism. *Environmental Ethics* 5(3), 237–243. https://doi.org/10.5840/enviroethics19835322

Tharakan, Koshy, Iype, Geevarghese V., & Afonso, A.V. (2011). Anthropocentrism and Ecocentrism: On the Metaphysical Debate in Environmental Ethics. *Jadavpur Journal of Philosophy* 21(2), 27–42. https://www.scribd.com/document/157609522/Anthropocentrism-and-Eco-centrism-On-the-Metaphysical-Debate-in-Environmental-Ethicsd-doc

Teelucksingh, Cheryl (2002). Spatiality and Environmental Justice in Parkdale (Toronto). *Ethnologies* 24(1), 119–141. https://doi.org/10.7202/006533arì

Tuan, Yi-Fu (1984). *Dominance and Affection: The Making of Pets.* Yale University Press.

Von Uexküll, Jakob (1934). Streifzüge durch die Umwelten von Tieren und Menschen: Ein Bilderbuch unsichtbarer Welten. J. Springer (2010, *A Foray into the Worlds of Animals and Humans: With a Theory of Meaning.* University of Minnesota Press).

Warren, Karen J. (1987). Feminism and Ecology: Making Connections. *Environmental Ethics* 9(1), 3–20. https://doi.org/10.5840/enviroethics19879113

Weitzenfeld, Adam, & Joy, Melanie (2014). An Overview of Anthropocentrism, Humanism, and Speciesism in Critical Animal Theory. *Counterpoints* 448, 3–27. www.jstor.org/stable/42982375

Whatmore, Sarah (2002). *Hybrid Geographies. Natures Cultures Spaces.* Oxford University Press.

Whatmore, Sarah (2006). Materialist Returns: Practising Cultural Geography in and for a More-than-Human World. *Cultural Geographies* 13(4), 600–609. https://doi.org/10.1191/1474474006cgj377oa

White, Lynn Jr. (1967). The Historical Roots of Our Ecologic Crisis. *Science* N.S. 155(3767), 1203–1207. www.jstor.com/stable/1720120

White, Richard J. (2015). Animal Geographies, Anarchist Praxis, and Critical Animal Studies. In Gillespie, Kathryn,& Collard, Rosemary-Claire (eds.), *Critical Animal Geographies: Politics, Intersections and Hierarchies in a Multispecies World* (pp. 19–35). Routledge.

White, Richard J., Springer, Simon,& de Souza, Marcelo Lopes. (eds.) (2016). *The Practice of Freedom. Anarchism, geography and the spirit of revolt.* Rowman and Littlefield.

Whitford, Philip, & Whitford, Kathryn (1951). Thoreau: Pioneer Ecologist and Conservationist. *The Scientific Monthly* 73(5), 291–296. https://www.jstor.org/stable/20438

Wilcox, Sharon, & Rutherford, Stephanie (eds.) (2018). *Historical Animal Geographies.* Routledge.

Williams, Raymond (1983). *Keywords: A Vocabulary of Culture and Society.* Flamingo.

Wolch, Jennifer, & Emel, Jody (1995). Guest Editorial: Bringing the Animals Back In. *Environment and Planning D* 13(6), 632–636. https://doi.org/10.1068/d130632

Wolch, Jennifer R., & Emel, Jody (eds.) (1998). *Animal Geographies: Place, Politics, and Identity in the Nature-Culture Borderlands.* Verso.

3 Assembling the toolkit

3.1 Making ecocritical geopolitics: research questions and analytical tools

The questions asked by ecocritical geopolitics are quite clear: what kind of understanding of the relationship between human and nonhuman is conveyed by popular culture? What kind of "environmental discourse" underlies it? And what categories of interpretation are used? Then there is the issue of "meaning": how, and where, is the meaning of a popular culture text produced/negotiated? To understand the real force of the discourse about the environment given in a text it is important not only to look at the author's position, but also at the genre of the text, the audience, and the circulation of the text.

The first step is to analyze the content, to understand whether there is a direct message about the need to respect what surrounds us, with its different components; and how the author's taken-for-granted world influences the text, that is, how the "geopolitical discourse" on which his/her interpretation of the world is reflected in the text. Next, it is important to take into account the genre of the text, which places the text in relation to a wider cultural system. Lastly, it must be remembered that both the author's message and taken-for-granted world can be grasped in different ways, depending on the audience and their own taken-for-granted world (because, as remarked by Umberto Eco, "the competence of the recipient is not necessarily that of the sender") (Eco 1983, 93). For these reasons, it is said that the meaning is "negotiated" between the production site (the author), the genre, and the receiving site (the audience). Finally, the circulation of the text must also be taken into account (has it been a success? Was it critically acclaimed?).

A good example of a single text whose meaning has been negotiated in very diverse ways is the *The Outlaw Josey Wales* (1976, Clint Eastwood), a film based on a novel by Forrest Carter (1972, *The Rebel Outlaw: Josey Wales*, later titled *Gone to Texas*). The movie is a western and follows the canons of the genre; it also makes use of some of the iconic actors of the genre in the 1970s, such as Clint Eastwood himself and Chief Dan George, made famous by his role in *Little Big Man* (Coyne 1998). Still, it is an anomalous western, because, instead of presenting (and glorifying) the westward march of a group of pioneers, it "describes the persecution and violent pursuit from Missouri to Texas of a gender and racially mixed band of the American oppressed" (Panay 2002, 32). Furthermore, in a speech that is "a filibuster by Eastwood's standards, and has strong political overtones: environmentalism, coexistence, distrust of faceless power structures" (Coyne 1998, 176), Josey, the main character, tells a Comanche leader called Ten Bears that his community will be able to coexist in peace with the Comanches, just as the wolf, the bear, the antelope do. In this way, the movie promotes an ecocentric message that stresses the importance of peaceful coexistence not only between humans of different cultures but also between humans and nonhuman animals.

For its ability to overturn the racial approach of the western genre and its typical ideology of conquest, the movie may be interpreted as "a critique of the frontier myth of American culture" (Panay 2002, 31) and be classified in the subgenre of the "revisionist western." Moreover, Josey is a rebel, fleeing the impositions of state law. This second theme is also fundamental from the director's point of view. Eastwood "claimed [*The Outlaw Josey Wales*] to be the most satisfying movie he ever made" and "always mentioned Vietnam and Watergate, and the kind of profound distrust that had developed toward government at the time" (Lowndes 2002, 249). For this reason, "the new community Josey and his friend forge in Texas" may be considered "emblematic of the multicultural consensus steadily evolving in the United States in the wake of Vietnam" (Coyne 1998, 178).

On the side of the viewer, however, interpretations can be conflicting: the antigovernment rhetoric expressed by the movie can be interpreted as "the representation through popular culture of an alternative American society" (Panay 2002, 36) but also as an "increasing distrust of national government" that may benefit the Right (Lowndes 2002, 237).

At the time of its release, the film was very successful, grossing $12,800,000; conversely, critics expressed discordant opinions, criticizing the excessive violence or appreciating its formal elegance (Coyne 1998). It was later selected by the National Film Preservation Board to be put in the National Film Registry due to its "cultural significance," and now, "Despite having been made after the golden age of westerns had ended, it is considered to be one of the best examples of the genre."[1]

Seen today, the film remains appreciable in its ecocentric approach, but it denounces the "geopolitical discourse" of its producers in terms of race and gender. Josey's community of oppressed people is made up of young and elderly women, and people of multiracial backgrounds. Still, the protagonist-leader is, unavoidably, a white male. All the other members of the group (women, non-white men, children, a dog) follow him without question. At the time, such a hierarchy of race and gender was part of the "taken-for-granted" world; now it appears to be outdated and "patriarchal" (Lowndes 2002).

As is clearly demonstrated, a text's meaning is negotiated in different sites (Rose 2016; Dittmer and Bos 2019): the site of production (What was in the author's mind when he/she wrote/composed that text? What discourse framed his/her way of understanding the world at that moment?); the content of the text itself (What is represented? To which genre does the text belong? Are there forms of intertextuality?); then there is the audience (Who is reading/watching/listening to the text? How do they interact with it?); lastly, there is circulation, which can be inhibited or promoted for different reasons and which certainly affects the impact and influence of the text and its contents. As written by Bryson (1991, quoted by Rose 2016, 56) "The power of the painting is there, in the thousands of gazes caught by its surface, and the resultant turning, and the shifting, the redirecting of the discursive flow." This observation applies, of course, to any kind of text.

Hence, even if the main theoretical reference point for ecocritical geopolitics remains popular geopolitics, which is a research approach that focuses on "the role of media and popular culture in shaping our understandings of the world around us" (Bos 2019, xiii), trying to answer its research questions requires a methodological "assemblage" that takes its instruments from different disciplinary fields and research traditions.

In this book, "text" is used to refer to any cultural product whether written, spoken, or visual. Even if the analytical tools used differ, the approach taken at the appropriate stages of research is generally the same. First of all, the content of the text — be it a single image, the lyrics of a song, a novel or a full movie — needs to be analyzed. For visual images, technical skills developed in visual research and film studies are useful. For written texts, narratology, semiotics, and literary criticism can be helpful. The study of the author's biography may also prove informative, in combination with the analysis of the interviews given at the time of the launch of the text, if the author is contemporary. In addition, knowledge of the literary or cinematographic genre of reference is indispensable to grasp intertextualities and recurring images and tropes, and meanings attributed to them. Should one want to analyze a large body of text, it is possible also to apply content analysis.

Then, since the specific interest of ecocritical geopolitics, as in popular geopolitics, is the connection between knowledge and power, in Foucauldian terms, the ultimate goal is discourse analysis. For discourse analysis, it is important to identify the interpretative categories with which the understanding of the world is built. In terms of relationship with the environment, it is necessary to refer to the different approaches highlighted in section 3.2. (anthropocentrism, biocentrism, etc.), and to some interpretative categories used by the geographical discourse, such as place, landscape, territory.

Lastly, audience studies should be made using tools typical of sociology and ethnography, such as surveys, focus groups, and netnography.

In order to offer researchers a point of departure, the chapter introduces some methodological notes. However, this book will mainly consider the first two sites of meaning, production and content, and its main focus will remain discourse analysis.

3.2 Analysis of the textual content: narrative structure, genre and composition

Analysis of textual content covers an examination of its narrative structure, genre and composition. Following Prince's definitions, a "narrative is the representation of at least two real or fictive events or situations in a time sequence, neither of which presupposes or entails the other" (Prince 1982, 4), regardless of the medium of representation. Narratology is the study of the form and functioning of narrative. Studying narrative structure means identifying the narrator (there is at least one narrator for each narrative), to see if he/she is an omniscient narrator, or is a character in the narrative (i.e. if he/she is extradiegetic, or diegetic), if he/she is more or less intrusive and so on. Then there are the "narratees," that is the characters, usually defined by their actions or words, and classified in accordance with generic categories of which their actions, words or feelings are illustrations, and in terms of the functions they fulfill (protagonists, heroes, villains and so on). The setting is the spatio-temporal complex where the narrative takes place. Altogether, a narrative has a "fabula" that is a fundamental scheme, with all the events in chronological order, and a plot. The plot is the story as it is in fact told, which may include time jumps (such as flashbacks) and other narrative artifices.

Placing the text within a specific genre, cinematic or literary, is sometimes immediate, sometimes less simple. It is based, in any case, on a generalization. The concept of genre, writes Todorov (1975), is borrowed from the natural sciences; and "scientific method proceeds [...] by deduction. We actually deal with a relatively limited number of cases, from them we deduce a general hypothesis, and we verify this hypothesis by other cases, correcting (or rejecting) it as need be. Whatever the number of phenomena (of literary works, in this case) studied, we are never justified in extrapolating universal laws from them; it is not the quantity of observations, but the logical coherence of a theory that finally matters" (Todorov 1975, 4)

Then, he adds: "Genres are precisely those relay points by which the work assumes a relation with the universe of literature." Generally speaking, "The genre system is a grid, and individual genres have boundaries, which are policed by the stakeholders who draw or map them" (Fletcher 2016, 3). Genres can thus be defined on the basis of many variables: the role of the hero (superior to the reader, by nature, as in myths, or by degrees, as in fairy tales, or inferior to the reader, as in irony), the verisimilitude, the spatio-temporal organization (the "chronotope"); for literary texts only, there is also the mode of fruition, which distinguishes works to be performed (drama) from works to be read (prose).

Returning to the example of *The Outlaw Josey Wales*, the movie can be defined as a western on the basis of the chronotope (it is set in the west of the North American continent, in the second part of the nineteenth century) and the qualities of the hero (a tall, slim American, skilled with a gun). However, because it sides with the Native Americans, not the settlers, it belongs to the subgenre of the "revisionist western." Placing a text within the genre to which it belongs helps to identify intertextuality frames, that is existing recorded narrative situations typical of the genre, which Umberto Eco (1983) calls "inferential walks."

On the basis of these "inferential walks," it becomes possible to attribute a recurring symbolic meaning to certain elements of the setting, or to the way the landscape is described/represented. "In broad terms, geography and genre are mutually constitutive" (Fletcher 2016, 1). For instance, to remain in the western genre, "The typical western film opens with the framing of a landscape" (Tompkins 1992, 69), from John Ford's "classical" era, to contemporary revisitations, such as *The Revenant* (2015, Alejandro González Iñárritu). It is a harsh and hostile landscape that challenges the man (male) to make it the object of conquest. It is a scarcely anthropized "big empty" (Engel 1994) that has yet to be completed by the arrival of those who can dominate and transform it. It invites the audience to admire the missionary task of the pioneers who strive to impose their own culture on such an unassimilated expanse of otherness; it is so unsuitable for whiteness that it becomes a monument to the narrative of the expansion towards the West, of the myth of the "frontier," of the foundation of the nation. At the same time, it is a "gender allegory" (Tompkins 1992, 82): identifying with the landscape, and with its harshness, with its being made of rock, allows one to express masculinity at the highest level.

Landscape, which is not an object, but the relation that is built in the text between the setting and the protagonists of the narrative, can acquire a symbolic dimension and even a capacity to act, becoming an "actant" (that is, an element that is valid for the place it occupies in the narrative and for the contribution it makes to carry that narrative forward). As such, it helps "to clarify the relationship between humans and the natural environment" (Peraldo 2012). In *Robinson Crusoe*, for instance, "The island, as seen through his eyes (eye-land), becomes a place where his [its] subjectivity and will are expressed (I-land). Of the way Crusoe reproduces his former spatial organization on the island, we might say that he turns it into just another city in an attempt to extract order from the apparent disorder of the natural world" (Peraldo 2012, 19).

In written texts, landscape can be described from the point of view of the protagonist (who performs the act of looking), or the omniscient / external narrator, who describes the scene within which the action takes place. Then come into play the richness of the description, the adjectives, the rhetorical figures, the color. As far as visual research is concerned, the composition of the image is engaged.

"Compositionality refers to the specific material qualities of an image or visual object" (Rose 2016, 25). So, compositional interpretation refers first of all to the content of the image (What does the image actually show?) and then to technical aspects. It pertains to color, spatial organization, and perspective (the *mise-en-scène*), focus, and point of view. With regard to the relationship between human protagonists and their surroundings, the "shot distance," which refers to how much of a figure is shown by a particular shot, is important. A shot can be an extreme long shot (where the figure is in the far distance), a long shot, or a full, three-quarter, medium, head-and-shoulders, or close-up shot (Monaco 2009). The long shot shows the subject from a distance, emphasizing his/her place, the position, and the relationship to the environment. The extreme long shot was introduced by the westerns of the 1950s and' '60s, "landscape movies" in which the grandeur of the setting and the smallness of the human figure highlight the challenge that the harshness of "nature" poses to the human being. Then there is the "bird's-eye shot" (or aerial shot), similar to the extreme long shot, but taken from above. In this case, the character still needs to be presented, but importance is given to the scenery. In the drama television series *True Detectives* (2014-2019, HBO), the show's photographic director, Nigel Bluck, "decided that, rather than a typical altitude of 200 to 500 feet, he would shoot from 1,000 feet — in part to amplify the abstract nature of the landscape, and in part because 'we wanted a point of view that belongs to something bigger than us'" (Sternbergh 2015). For this reason, in *True Detective* the landscape may be considered an "actant" (Amato 2015). Close-up shots reveal the details of the subject and highlight the emotions of a character. Medium shots emphasize the subject, but still show him or her as part of the surrounding environment. Then there is the "cowboy shot," larger than a medium shot, and smaller than a full shot. The character is seen from the hip up, in a superior position, exhibiting a confident, sometimes heroic, attitude. The "cowboy shot" was introduced in westerns to frame the gunslinger's gun.

The use of color tones can also communicate something about the relationship between human beings and the environment. In the TV series *The Walking Dead* (2010–, AMC), each season shows a greener setting, demonstrating how "nature" is continuing its course, regardless of the near extinction of human beings. Conversely, Furiosa, the female lead in *Mad Max Fury Road* (2015, George Miller), crosses a monotonously yellow desert to indicate the impact of desertification and climate change on the future world.

Instead of considering a single text, or a genre with a qualitative approach, content analysis examines a large number of images or written texts using a quantitative approach. It consists in encoding some elements present in the texts and verifying their occurrence within a given textual corpus (a magazine, a certain number of films shot within a given period of time, etc.). In analyzing the setting of post-apocalyptic films shot between 1950 and 2010, for example, content analysis can be made by counting how many times an adverse climatic event is represented, how it is represented and its nature (hurricane, flood, drought, etc.) One of the merits of content analysis is its replicability, though it is clearly necessary for the coding process to be carried out as precisely as possible for this to happen.

3.3 Territory, place, landscape: clarifying some geographical notions

In the toolkit of ecocritical geopolitics, in addition to narratology, semiotics, and visual research, there are a number of essential "tools" for geographical analysis, such as the notion of landscape, territory, and place.

"Prevailing ideas about 'nature' have implications for the treatment of certain categories of humans who are 'natured' and have certainly impacted non-human species of animals, plants and their contexts" (Cudworth 2005, 43). The portion of space, shared by different "living beings" (whatever definition, more or less inclusive, is given to this expressions), that Cudworth calls "context" is the object of transformation processes related to human activities. Geographers call such human-transformed context "territory." More specifically, following Raffestin (1980), territory is the result of an action carried out by an actor on space. By appropriating a space, concretely or abstractly (e.g. through representation), the actor territorializes this space, constantly reorganizing its pre-existing conditions. The process through which human beings modify the surface of the Earth, turning physical space into a social artefact ("territory"), is called "territorialization". Territorialization follows three different steps: denomination, that is the symbolic elaboration of space, through its semiotic signification; reification, that is its physical transformation, through activities such as agricultural production, water control, and setting; and structuration, that is the control over the political, social, and administrative organization of space (Turco 2010).

According to the Italian geographer Angelo Turco (2010), territoriality is the "fundamental geographical quality of the world," while environment, place, and landscape are simply "configurations of territoriality." Place is a device for establishing a relationship with the space; it is a portion of space (of territory) made meaningful to someone. Place is therefore a fundamental interpretative category, constructed and reconstructed to give meaning to the space in which one moves and acts; it is, in a certain sense, the symbolic context elaborated by human beings to act in the world. Space perceived as a place may become an affective object, a matter of love, of *topophilia* (Tuan 1974), or of fear or anxiety, of *topophobia* (Relph 1996). Place is not just a structure of feeling. It is a portion of space that someone may claim to belong to, and even to possess. Place, once delimited, establishes who is inside and who is outside, and becomes a device to define who lives on the outside as someone who is "different": in this way, place takes on a further meaning, that of an instrument of identity differentiation and geographical dualism. People "make places" out of spaces, but they may make them on the basis of conflicting interpretations and images. This also has consequences in relation to environmental management, as is made clear, for example, by the Muir-Pinchot dispute over the O'Shaughnessy Dam in the Hetch Hetchy Valley.

The concept of landscape is connected to the concept of place. Landscape is a symbolic reinterpretation of the visual (but also olfactory and sensory) relationship between human beings and their context (their territory, and their place). Therefore, landscape, like place, has a subjective dimension. As highlighted by Denis Cosgrove (1984), it is a way of seeing/representing the world that has its own history, and for this reason needs to be understood as part of the broader history of the economy and society. As such, landscape is the result of the sum of different interpretations and derives from an intricate web of intertextual references. In literature, and in cinema, the landscape differs from the setting in that it is not the site where the action takes place, but a representation of that site, and of the perception of the relations between that site and the human character (or nonhuman animal) of the action (dell'Agnese 2016). If popular culture constitutes "the vehicle of the representations that a society gives of itself" (Aumont, Bergala, and Vernet 1983), then landscape becomes the representation of the relationship that society has, or thinks it has, or would not like to have, with its own territorial context (with its environment, with its places). In its articulation, therefore, landscape makes it possible to convey environmental themes, or to enrich the text with a social/environmental warning that goes beyond the plot and can sometimes be read independently of it.

Environment is the territorial shape of "nature," the "natural" dimension of the context (Turco 2010). Still, territory is "produced" by recursive processes dependent on the relationship between nature and culture in different spatial contexts (Raffestin, 2012). In this perspective, the opposition of nature and culture does not exist precisely because the "nature" that is transmitted from one state of nature to another is strongly marked by earlier cultures (and vice versa).

3.4 Discourse analysis

Discourse analysis is a qualitative research approach, closely linked to the concept of discourse introduced above and to Michel Foucault's philosophy. It is therefore based on the idea that the production of knowledge is a form of power. As noted by Rose (2016, 189), "Discourse, Foucault says, is powerful, but it is powerful in a particular way. It is powerful, says Foucault, because it is productive. Discourse disciplines subjects into certain ways of thinking and acting, but this is not simply repressive; it does not impose rules for thought and behavior on a pre-existing human agent. Instead, human subjects are produced through discourses." And then she adds "knowledge and power are imbricated one in the other, not only because all knowledge is discursive and all discourse is saturated with power, but because the most powerful discourses, in terms of the productiveness of their social effects, depend on assumptions and claims that their knowledge is true. The particular grounds on which truth is claimed — and these shift — historically constitute what Foucault called a regime of truth".

Initially developed within linguistics, discourse analysis then saw the possibility of being applied also to visual research (Rose 2016), and to many other forms of communication (and even to physical objects in everyday life) (Dittmer and Bos 2019). So, beyond language and image, discourse analysis can be applied also to "toys, monuments, films, sounds, etc.," which together may combine to make a meaning" ("multimodal analysis") (Berger 2017, 9). Following the example proposed by Dittmer and Bos (2019), for instance, it is possible to analyze the discourse on race in popular culture for children, focusing on *Tarzan*, and to this end examine old films, comics, and even toys. When making discourse analysis, it is important to look not only at the plot, or the fabula, but also at the details, because "everything in a text such as a play, a film, a commercial, or a photograph, is important" (Berger 2017, 81).

Discourse analysis starts with a research question and continues with the choice of source: what to examine to answer that question? Returning to the example of the discourse on race in popular culture for children, if Tarzan, the story of a little white boy, raised by apes in Africa, is a pretty obvious choice, a cartoon like *Tom and Jerry* (1940, William Hanna and Joseph Barbera), whose protagonists are a cat called Tom and a mouse called Jerry, is not. Looking at the details, however, one can notice that the only human being that appears in the cartoon is a maid, and that that maid (of whom we just see her calves) is black. This example also helps to explain how intertextuality, both within and across genres, activates, as suggested by the Italian semiologist Umberto Eco (1983) the "encyclopedic competence" of the audience, reinforcing and validating the "interpretative repertoires" that are available in a particular cultural setting, the discourse and its regime of truth. The (fat) calves of *Tom and Jerry's* maid are in fact an intertextual quotation from other female characters from US popular culture, all dark-skinned and overweight, serving the (white) protagonists of the narrative. In this way, the so called "Mammy" stereotype (Wallace-Sanders 2008) and a given racial hierarchy as a regime of truth are reinforced.

The texts examined need not necessarily be of the same type; as noted by Rose (2016, 196), "Some of the most interesting discourse analyses are interesting precisely because they bring together, in convincing ways, material that had previously been seen as quite unrelated." In fact, a novel like *Uncle Tom's Cabin* (1852, Harriet Beecher Stowe), a movie like *Gone with the Wind* (1939, Victor Fleming), and a cartoon like *Tom and Jerry* seem to have very little in common, but they are all texts that reinforce the same stereotype. Going one step further, we may move on to the "popular geopolitics of the everyday" (Dittmer and Bos 2019), and mention the Italian fast-food chain called "Mama Burger," which has as its emblem the face of a Mammy.

3.5 What about the audience?

The role of the audience is important from different points of view.

First of all, it can influence the content of the text. Generally speaking, the production can try to be more or less "commercial," adapting the content to the taste of the "general public." More specifically, in the case of serial production, the audience's expectations can push the production to modify the plot and the characters. This already happened in the late nineteenth century in the case of the serial novel. *The Mysteries of Paris* (Eugène Sue) was published serially in 90 parts from June 1842 until 15 October 1843, making it one of the first serial novels published in France. As Eco (1983) points out, the novel was initially written to entertain the cultured public with some spicy events of picturesque misery, but it was instead welcomed by the proletarian public as a description of their condition. Thus, the author, who noticed this reception, continued to write it for the proletariat, filling it with social democratic moralities to convince these "dangerous" classes to hold back their despair. In similar fashion, contemporary television series and transmedia franchises often adapt characters and plots to preferences expressed by the public. Thus, *The Walking Dead* (2010–) was initially accused of presenting a "regressive conception of gender" and extending its stereotypes to promote "racial coding"(Sugg 2015); "the entire narrative [was] framed by the white male supremacy of the main character" while characters of color and women were cast in subordinate positions (Baldwin and McCarthy 2013, 73). During the seasons that followed, however, the production tried to remedy this by emphasizing female leadership and increasing the presence of black characters.

Second, the public plays an important role from a quantitative point of view: the greater the number of people who see or read a given text, the greater its impact will be. In this case, sales figures, box office success, number of copies sold, but also the circulation between different media (called transmediality), and even the number of times a product has been illegally downloaded from the web are useful.

Finally, and perhaps above all, it is important to understand how the audience interprets the text, or rather, how viewers and readers, in different conditions, endowed with different skills and competences, negotiate the meaning. It has already been pointed out that the competence and interpretative categories of those who read or look at the text can be very different from those who produced it. As Umberto Eco writes (1983), the readers (or viewers) have their own "encyclopedic competence" at their disposal, a "world of references" through which to make sense of the content of the text. Since this world of reference is a cultural product, just as the point of view of the production is culturally produced, the meaning is the result of constant negotiation.

To investigate this last point, there are several research methods available, qualitative and quantitative, which involve both direct contact with the reader/spectator and the investigation of opinions expressed on the web. Among the more traditional methods, surveys (i.e. questionnaire surveys), interviews and focus groups are all widely proven research methods in the sociological field. Questionnaire research allows a good amount of data to be collected, but, in order to be significant, it must be aimed at a representative sample. Interviews can be based on structured questions or be unstructured. It also possible to conduct photo-elicitation interviews, in order to investigate the reaction caused by a single image, or by a group of images. In both cases, interviews make it possible to collect detailed qualitative information. Eventually, to do research with a focus group means to assemble a group of people and make them discuss the meaning of the text or group of texts that are the subject of the research.

Besides stimulating the expression of an opinion by the public, with questionnaires, interviews, and focus groups, it is also possible to analyze opinions already expressed on the web by investigating databases using specific research approaches, a technique called netnography and sentiment analysis. Netnography is ethnographic research through the internet (Kozinets 2002). Specifically, doing netnography means giving value to comments and narratives left on the web, in social networks, forum discussion or blogs. These new spaces of communication are rich in stories, relationships, opinions, and comments. Netnography is a qualitative approach that can be used to explore comments left by readers on social networks and on databases, such as Community Reviews on Goodreads,[2] or a movie's user reviews on Rotten Tomatoes[3] and IMDb.[4] Specifically, IMDb "is not only a database but also a social network whose users share their opinions and knowledge, contributing content to the largest digital collection of data not only on films but also on television programs and video games" (Canet Centellas et al. 2016, 152). It is possible to carry out "opinion mining" on the same sources, also with a quantitative research approach, called "sentiment analysis," which can help in handling a "huge volume of opinionated data recorded in digital forms" (Liu 2012, 5). Sentiment analysis makes it possible to combine the analysis of comments relevant to the research question with the influence of those who make them (measured in terms of followers), and thus the potential ability to exert an impact on other viewers/readers.

Notes

1 https://civilwar.vt.edu/the-outlaw-josey-wales-1976/
2 www.goodreads.com
3 www.rottentomatoes.com
4 www.imdb.com

Bibliography

Amato, Fabio (2015). La Louisiana di True Detective. Attori e attanti geografici in una serie televisiva. *Acoma* 9, 114–123.

Aumont, Jacques, Bergala, Alain, Marie, Michel & Vernet, Marc (2008). *Esthétique du film*. Nathan. (1992, *Aesthetics of film*, University of Texas Press.)

Baldwin, Martina, & McCarthy, Mark (2013). Same as it ever was: Savior narratives and the logics of survival in The Walking Dead. In Balaji, Murali (ed.), *Thinking Dead: What the Zombie Apocalypse Means* (pp. 75–87). Lexington Books.

Berger, Arthur Asa (2017). *Applied Discourse Analysis: Popular Culture, Media, and Everyday Life*. Springer.

Bos, Daniel (2019). Preface to the Second Edition. In Dittmer, Jason, & Bos, Daniel (2019). *Popular Culture, Geopolitics, and Identity* (pp. xiii–xiv). Rowman & Littlefield.

Bryson, Norman (1991). Semiology and Visual Interpretation. In Bryson, Norman, Holly, Michael Ann, & Moxey, Keith P.F. (eds.), *Visual Theory: Painting and Interpretation* (pp. 61–73). Polity Press.

Canet Centellas, Javier Fernando, Valero Navarro, Miguel Angel & Codina Bonilla, Lluis (2016). Quantitative Approaches for Evaluating the Influence of Films Using the IMDb Database. *Communication & Society* 29(2), 151–172. [https://doi.org/10.15581/003.29.2.151-172]

Cosgrove, Denis E. (1984). *Social Formation and Symbolic Landscape*. Croom Helm.

Coyne, Michael (1998). *The Crowded Prairie. American National Identity in the Hollywood Western*. I.B.Tauris.

Cudworth, Erika (2005). *Developing Ecofeminist Theory: The Complexity of Difference*. Springer.

dell'Agnese, Elena (2016). Il paesaggio come metafora: l'approccio della Critical Geopolitics. In Frisina, Annalisa (ed.), *Metodi visuali di ricerca sociale* (pp. 107–123). Il Mulino.

Dittmer, Jason, & Bos, Daniel (2019). *Popular Culture, Geopolitics, and Identity*. Rowman & Littlefield.

Eco, Umberto (1983). *Lector in fabula. La cooperazione interpretativa nei testi narrativi*. Bompiani.

Engel, Leonard (1994). *The Big Empty: Essays on Western Landscapes as Narrative*. University of New Mexico Press.

Fletcher, Lisa (2016). Introduction: Space, Place, and Popular Fiction. In Fletcher, Lisa (ed.), *Popular Fiction and Spatiality. Reading Genre Settings* (pp. 1–7). Palgrave Macmillan.

Kozinets, Robert V. (2002). The Field Behind the Screen: Using Netnography for Marketing Research in Online Communities. *Journal of Marketing Research* 39(1), 61–72. https://doi.org/10.1509/jmkr.39.1.61.18935

Lefebvre, Henri (1974). *La production de l'espace*. Anthropos (1991, *The Production of Space*. Monoskop).

Liu, Bing (2012). *Sentiment Analysis and Opinion Mining*. Morgan & Claypool Publishers.

Lowenthal, David (2000). Nature and Morality from George Perkins Marsh to the Millennium. *Journal of Historical Geography* 26(1), 3–23. https://doi.org/10.1006/jhge.1999.0188

Lowndes, Joseph (2002). Unstable Antistatism: The Left, the Right, and "The Outlaw Josey Wales." *International Journal of Politics, Culture, and Society* 16(2), 237–253. www.jstor.com/stable/20020161

Monaco, James (2009). *How to Read a Film: Movies, Media, and Beyond*. Oxford University Press.

Panay, Andrew (2002). The Outlaw Josey Wales and the Frontier Myth of Development. *Media Education Journal* 32, 31–38.

Peraldo, Emmanuelle (2012). 'Two Broad Shining Eyes': Optic Impressions and Landscape in Robinson Crusoe. *Digital Defoe* 4(1), 17–30.

Prince, Gerald (1982). *Narratology. The Form and Functioning of Narrative*. Indiana University.

Raffestin, Claude (1980). *Pour une géographie du pouvoir*. LITEC.

Raffestin, Claude (2012). Space, Territory, and Territoriality. *Environment and Planning D: Society and Space* 30(1), 121–141. https://doi.org/10.1068/d21311

Relph, Edward (1996). Place. In Douglas, Ian, Huggett, Richard, & Robinson, Mike (eds.), *Companion Encyclopedia of Geography. The Environment and Humankind* (pp. 906–922). Routledge.

Rose, Gillian (2016). *Visual Methodologies: An Introduction to Researching with Visual Materials*. Sage.

Saunders, Robert A., & Strukov, Vlad (eds.) (2018). *Popular Geopolitics: Plotting an Evolving Interdiscipline*. Routledge.

Sternbergh, Adam (2015). *True Detective* and the Changing, Divisive Nature of the Aerial Shot. *Vulture*. www.vulture.com/2015/07/true-detective-aerial-shot.html

Sugg, Katherine (2015). The Walking Dead: Late Liberalism and Masculine Subjection in Apocalypse Fictions. *Journal of American Studies* 49(4), 793–811. https://doi.org/10.1017/S0021875815001723

Todorov, Tzvetan (1975). *The Fantastic: A Structural Approach to a Literary Genre*. Cornell University Press.

Tompkins, Jane (1992). *West of Everything: The Inner Life of Westerns*. Oxford University Press.

Tuan, Yi-Fu (1974). *Topophilia. A Study of Environmental Perception, Attitudes And Values*. Prentice-Hall.

Turco, Angelo (2010). *Configurazioni della territorialità*. Franco Angeli.

Wallace-Sanders, Kimberly (2008). *Mammy: A Century of Race, Gender, and Southern Memory*. University of Michigan Press.

2

Landscapes and fears

Discourse about the environment (and unavoidably also about race and gender) in dystopian texts and post-apocalyptic narratives

4 Re-visioning the future

4.1 Popular culture and landscapes of fear

Landscapes of Fear is the title of a book by Yi-Fu Tuan, published in 1979, in which the Sino-US geographer demonstrates how our fears are influenced by the historical moment we inhabit and in which we socialize. Some fears are subjective, others arise from a specific reason, but they are not permanent moods linked to unalterable segments of tangible reality. We are "socialized" to fear something because we receive threats and warnings about it, whether the danger is real or presumed. So, while retaining our fears, we also adopt other, socially formulated, ones (Claeys 2016, 16).

For this reason, to understand fears we must approach them from both the individual and the group perspective and place them in a historical frame. The need to contextualize fear is due to the changes in the social system, technological skills, and modes of relationship between human beings and the environment. In some cases, the dangers do not change; what changes is the way they are perceived. So, even though "more Americans die from accidents in the bathtubs than from attacks by terrorists,"[1] nobody fears bathtubs, they fear terrorists.

The perception of fear is linked to the representation of what constitutes a danger: we learn to be afraid of something because it is described to us as threatening. In this sense, the film *The Village* (2004, M. Night Shyamalan) provides an excellent narrative about the making of fear (Sánchez-Escalonilla 2010). In the village of the title, a small community lives surrounded by woods and disconnected from the outside world. It is impossible to leave the village and break the seclusion because ferocious creatures of superhuman nature are said to live in the woods. In reality, there are no monstrous creatures. The wood is built up as a "landscape of fear" by the village elders to prevent young people from coming into contact with the threats posed by external contemporary society. The movie places its viewers "in the position of citizens who have been deceived by their leaders" (Collier 2008). At the same time, the film shows how the elderly are afraid of something (in this case, the threats of contemporary society) and use representation to teach young people how to avoid what they fear. Studying the relationship between fear and representation, therefore, helps us to understand, on the one hand, what we are afraid of, and, on the other, what we teach others to fear.

A tale about teaching young people what to be afraid of and how to behave accordingly is not a Hollywood invention. Medieval fairy tales acted as "cautionary tales," making the wood or the castle a "landscape of fear" and teaching children and adolescents not to stray too far from the family settlement. Certain images and metaphors have been used regularly to "scare" people and thus direct them towards "correct" behaviors. One of them is the apocalypse, which does not necessarily mean (only) the physical destruction of the world; it refers to a "discontinuity" and the subsequent destabilization of the ontological and epistemological experience. Other narrative strategies, such as dystopian stories, have become increasingly popular as a warning about the social and political dangers that could cloud the future of the planet.

Dystopian and apocalyptic narratives serve as contemporary cautionary tales, warning against what is perceived as a danger. In this perspective, they may be considered "as a window" into the cultural anxieties of the time (Gergan, Smith, and Vasudevan 2018, 2). They therefore express the prevailing fears at a given historical time. They also show how popular culture communicates these fears to people.

If we take the relationship between children and the forest as an example, we see that at one time children were taught to fear its creatures. *Little Red Riding Hood* warns children not to enter the wood so as not to meet the wolf. Today, there are fears that the forest will disappear. By seeing, in a post-apocalyptic movie, a forest-free future, children may learn how to behave so that their children, and the children of their children, will have woods to walk in. Alternatively, a dystopian future where only the rich can access that forest, while the poor live in a devastated environment, may teach viewers how to promote environmental justice. If wolves and deer live in that forest, children will need to ensure that the animals have no reason to fear being hunted.

4.2 Dystopian texts and post-apocalyptic stories

Setting clear boundaries on the dystopian narrative as a "narrative strategy" and unambiguously distinguishing it from apocalyptic and post-apocalyptic narratives are not easy. Generally speaking, the term "utopia" and its various derivatives (eutopia, dystopia, utopian satire, anti-utopia) are defined as "a species of prose fiction that describes in some detail a non-existent society located in time and space" (Tower Sargent 1976, 275). If the representation of an ideal world, so perfect as not to exist, is defined as a utopia, a dystopia is, in opposition, a bleak place where, due to an illiberal political system, or to health, environmental or technological disasters, or to an environmental crisis, human society has plunged into an undesirable state. A dystopia is usually placed in a future world. It can be the consequence of a political regime, which may appear utopian to those who impose it, but is unbearable to those who suffer it. Alternatively, it depicts a place where some contemporary tendencies have been taken to extreme consequences, leading to a highly unpleasant situation. In this form of classical dystopia, the author makes a point of describing the horrors of life if current trends continue, "if This Goes On" (Ketterer 1989, 212).

A post-apocalyptic narrative, on the other hand, necessarily involves a break-up moment. It represents the post-anthropic setting of the post-disaster, highlighting the transience of the forms of occupation of human space and the precarious balance that binds contemporary societies to their environmental context. The myth of the apocalypse, acting as a call to repentance and redemption (Yanarella 2001), is perfectly suitable to emphasize the potential risks posed by the lack of respect towards the environment for the future of the planet and humanity (Killingsworth and Palmer 1996). Moreover, it introduces the possibility of reflecting on the possible consequences that a breakdown in order can exert on the relations between individuals, causing anarchy and a return to barbarism, or, on the contrary, the loss of democracy and the establishment of an intolerable political and social system. In this case, the post-apocalyptic society reproduces a world so negative as to be dystopian.

To some critics, the definition of dystopia should exclude texts that refer to "impossible" events. In this perspective, a dystopia should represent only "feasible negative visions of social and political development, cast principally in fictional form" (Claeys 2010, 109). By "feasible," the implication is that extraordinary or utterly unrealistic elements should be excluded from the narrative. Such a distinction makes it possible to draw a clear line between dystopia and sci-fi, but it also narrows the field of analysis considerably. For instance, if taken to the letter, it excludes all the texts adopting the narrative stratagem of the "long sleep," from *When the Sleeper Wakes* (a serialized story by H.G. Wells published first in *The Graphic*, then republished as a novel in 1899 and then, in a new version, in 1910 with the title *The Sleeper Awakes*)[2] to the movie *Sleeper* (1973, Woody Allen), or *Idiocracy* (2006, Mike Judge). Moreover, some texts represent a world made dystopian by the advent of a phenomenon that is not "feasible" (in the television series *Colony*, 2016–18 USA Network, for example, the world is subject to totally illiberal — and therefore dystopian — forms of government following the "arrival" of mysterious, and alien, "hosts").

Kunkel (2008, 90) suggests a distinction between apocalyptic and post-apocalyptic texts and dystopias. In his opinion, these categories refer to different and sometimes even opposite scenarios. The end of the world, or apocalypse, brings the collapse of order as a typical consequence, while a dystopia often prefigures a sinister perfection of order. A clear-cut distinction, however, is problematic also in this case; post-apocalyptic stories often present chaos, because "without an orderly, strong, centralized authority, people are reduced to base instincts and criminal behavior. People loot, murder, rape, and cannibalize" (Booth 2015, 18). After a few years, however, in response to the instability of the "end of the world," a despotic and unjust "dystopian" order may be reconstituted, as in *The Handmaid's Tale* (a novel by Margaret Atwood, 1985, adapted into a movie in 1990 by Volker Schlöndorff, and into a TV series,

2017–…), and in much of Young Adult Fiction, such as *The Hunger Games* trilogy (2008-10, Suzanne Collins) and the *Divergent* saga (2011-13, Veronica Roth). In a more inclusive definition, we can label the representation of a dark future as dystopian, whatever its cause, and accept the fact that it often overlaps with a post-apocalyptic narrative, and sometimes with sci-fi.

Three main forms of dystopian narratives can be identified, depending on the anxieties highlighted by the plot: political dystopia, environmental dystopia, and technological dystopia (Claeys 2016, 5).

All three were already formulated, albeit with different nuances, at the turn of the nineteenth century, with the novels of H.G. Wells (Hillegas 1967). However, they blossomed at different times during the twentieth century with the rise of political and economic themes and the representation of oppressive systems. Environmental and Malthusian themes, linked to the fear of excessive population growth and lack of resources, then became relevant. Technological dystopia, linked to the growing inability of human beings to control machines or the digital universe, is currently very successful. Because it focuses on human events, dystopia is usually anthropocentric, even though it may include considerations on issues of environmental justice and may embrace an ecofeminist discourse.

The apocalypse is a metaphor that dates back to ancient times, to the Bible and the *Epic of Gilgamesh*. In popular culture, it has enjoyed several moments of success, related first to the fear of plagues, then of nuclear disasters, and finally of unstoppable climate change. More recently, the post-apocalyptic/horror theme has become fashionable, where the human species undergoes alteration (often due to human genetic manipulation) and is (almost totally) transformed into something else (often zombies). With regard to environmental discourse, the post-apocalyptic narrative offers many intriguing elements for analysis: those connected with the plot (the reason for the catastrophe) and those related to the setting and landscape (the future of the planet).

4.3 Increasingly successful narratives

Even if they often overlap, dystopian and apocalyptic or post-apocalyptic narratives have different origins. In Western literature, one of the first novels to describe the events during a possible end of the world is probably *Le Dernier Homme* (1805) by the French author Jean-Baptiste Cousin de Grainville (published in England as *The Last Man* the following year). In 1816, Lord Byron wrote a short poem called *Darkness* in which the world is a lifeless "void," the oceans and rivers stand still in the dark, and even the moon is dead. In the ten years that followed, the topic became popular among Romantic authors, and there was a small stream of publications on the subject, including Mary Shelley's novel also titled *The Last Man* (1826).

Apart from *Gulliver's Travels* by Jonathan Swift (1726), which is dystopian in some parts, the first political and social dystopias date back to the early nineteenth century. Swift, in his most famous work, describes a series of strange, and sometimes very unpleasant, places and peoples through the rhetorical stratagem of travel in order to pillory some aspects of his contemporary British society. *A Sojourn in the City of Amalgamation, in the Year of Our Lord, 19—* written by Jerome B. Holgate and published in 1835 under the pseudonym of Oliver Bolokitten, was an early dystopian novel. It represents a future city dominated by "racial fusion," with the consequent moral and economic degeneration of society (Lemire 2002). Among the first significant dystopian satires, there is also the French novel *Le monde tel qu'il sera* (1846, by Émile Souvestre, English title *The World As It Shall Be*). The novel is about a French married couple who, accompanied by a man called John Progress, arrive in the year 3000 to discover that the world has become hyper-technological, but ruthless towards the most vulnerable. Another novel, in some way a forerunner of contemporary dystopias, is *Erewhon* (1872, by Samuel Butler). In a country called Erewhon (an anagram of Nowhere), everything works in reverse, and there are no machines because the inhabitants do not trust them and are afraid that they may develop a consciousness. These first novels are all travelogues, in space or in time, and their "narrative standpoint" is external to the dystopia. The protagonist visits the dystopian place, is shocked by its characteristics, sometimes falls in love with a local woman, and then succeeds in going away.

Toward the end of the nineteenth century, other novels and short stories with post-apocalyptic themes were published. *After London* (1885, Richard Jefferies), for instance, portrays London's renaturalization after an unspecified "ecological" catastrophe (Claeys 2010). Then, the first H.G. Wells' stories shuffled the themes: they all offer appalling representations of the future (Hillegas 1967) but some are "travelogues" in time, like *The Time Machine* (1895) and *When The Sleeper Wakes* (1899); others, like *The War of the Worlds* (1897), mix sci-fi and dystopia.

Dystopia, as a literary genre, became fully established during the twentieth century. During the first half, some novels were published that went on to become classics of the genre (*We*, 1924, by the Russian Yevgeny Zamyatin; *Brave New World*, 1932, by Aldous Huxley; *Nineteen Eighty-Four*, by George Orwell, 1949; *Fahrenheit 451*, 1953, by Ray Bradbury). They all describe a future in which an all-encompassing regime erases the freedom and agency of the individual. The plots no longer take the form of a travelogue, and the "narrative standpoint" is internal: the (male) protagonist is often a member of the system who, at a certain point, succeeds in regarding the situation with a critical eye.

Post-apocalyptic fiction enjoyed a new phase of popularity in the second half of the century, when a series of dramatic narratives expressed post-nuclear fears and environmental anxieties (*Pebble in the Sky*, 1950, by Isaac Asimov; *Level 7*, 1959, by Mordecai Roshwald; *The Chrysalids*, 1955, by John Wyndham, and *A Canticle for Leibowitz*, 1959, by Walter Miller). Generally speaking, "prior to the Second World War the most arresting film imagery depicting planetary disaster was either cosmic or natural in origin" (Broderick 1993, 363), but now "Hiroshima unleashed the possibility of man's annihilation at the hands of his own technology" (Brereton 2005, 143). This awareness provoked the first "green" apocalypses, in which the narrative becomes "one of human, not divine history" (F. Buell 2003, 13).

Malthusian anxieties are merged with a shortage of resources against the backdrop of the energy crisis. This is properly synthesized in Harry Harrison's novel, *Make Room, Make Room!* (1966, later adapted in the film *Soylent Green*, 1973, by Richard Fleischer). The story is about a near future marked by uncontrolled population growth and environmental devastation, when the rich still have fresh food, while all the others eat tasteless artificial preparations distributed by a single large corporation (Soylent). So, in addition to sounding a conservationist alarm about the future of the planet, the text launches a message of environmental justice, highlighting how the deterioration of the environment risks hitting only the weakest, leaving the privileges of others unaltered.

With overpopulation, pollution, and scarcity, the climate issue is touched upon by popular culture quite early, even if, at the time, the prospect of human-induced climate change was not evident. In 1941, George R. Stewart published *Storm*. This novel follows the path of a cyclone that forms in Japan and arrives in California, causing damage, floods, and deaths. In 1948, the sequel, *Fire*, tells the drama of a massive forest fire in California, ignited by lightning but propagated by drought, and the resulting environmental damages. In this regard, it is worth mentioning also J.G. Ballard's climate novels (Dini 2019) that each refer to a climatic cataclysm: a steady wind (*The Wind From Nowhere*, 1961), the unstoppable rise of the sea level (*The Drowned World*, 1962), and a terrible drought (The Burning World, 1964, republished as *The Drought*, 1965). Later, sci-fi movies started to use "climate change" images to emphasize their dramatic setting and increase the "cognitive estrangement" typical of the genre (Suvin 1977). Thus, in *Blade Runner* (1982, Ridley Scott), the sky over Los Angeles is always dark due to persistent rain, and in *A.I. Artificial Intelligence* (2001, Steven Spielberg) Manhattan has been submerged owing to the rise in sea level. Climate change returns as a protagonist in two movies from the 1990s, *Waterworld* (1995, by Kevin Reynolds) and *Tank Girl* (1995, by Rachel Talalay). In the first, the world is covered with water, in the second it is a desert. In both cases, there is a struggle for drinkable water. *Tank Girl* is noteworthy because the protagonist is a teenager who somehow anticipates the Young Adult Fiction heroines and the film is now considered a precursor of feminist themes (Zaslow 2009).

In the first two decades of the 2000s, climate change has become a telegenic phenomenon, and films on the subject have multiplied, giving rise to a new subgenre (cli-fi). This includes films in which the world of the future is submerged (*Lost City Raiders*, 2008, Jean de Segonzac), lacking water (*Interstellar*, 2014, Christopher Nolan; *Mad Max Fury Road* (2015, George Miller) or frozen (*The Day After Tomorrow*, 2004, Roland Emmerich; *Snowpiercer*, 2013, Bong Joon-ho). There are also many movies set in the present, where extreme weather events connected with climate change create catastrophic results, like *NYC: Tornado Terror* (2008, Tibor Takacs), the TV-movie *The Christmas Twister* (2012, Peter Sullivan), and *Geostorm* (2017, Dean Devlin). In US popular culture, the representation of extreme weather events is so clichéd that it also stimulates parodies, such as the super-trash film series (*Sharknado*, 2013-2018, Anthony C. Ferrante), which "tells the story of a giant global warming-induced tornado supercluster that can lift sharks *en masse*" (Leikam 2017, 31). Marketed as a B-movie, *Sharknado* is, in Leikam's opinion (2017, 31), a "self-ironic rendering of the staging of extreme weather as a media spectacle."

Another very popular subject is technological dystopia. In the *Terminator* franchise (first film, 1984, by James Cameron), the world is dominated by machines. In *Minority Report* (2002, by Steven Spielberg, based on a short story by Philip K. Dick, 1956), thanks to the predictive abilities of a small group of mutants and the technological ability to read their predictions, it is possible to prevent crime from happening. However, as the accuracy of the forecasts is not infallible, preventive intervention can be mistaken. Some episodes of the British television series *Black Mirror* (2011–2019) show the humanity of the future entirely conditioned by the "likes" collected on social media. Parallels can be found in the novel *The Circle* (2013) by Dave Eggers, whose theme is the power wielded by the giants of the internet, in a world where, in the name of "transparency," privacy has become a crime.

Other fears, such as the fear of losing fertility, or of genetic modification, are additions to the repertoire (plagues, totalitarianism, capitalism, invasions) for the creation of story lines that are becoming increasingly widespread and popular. Alongside texts aimed at young adults — often mixing apocalypses, zombies, and parodies — the post-apocalypse is also found in high-level works like Corman McCarthy's award-winning novel *The Road* (2006). In 2009, Kunkel commented: "Every other month seems to bring the publication of at least one new so-called literary novel on dystopian or apocalyptic themes and the release of at least one similarly themed movie" (Kunkel 2008, 89). The trend has not stopped yet. Besides, there are not only novels and movies. Beyond the literary form, dystopian and post-apocalyptic storytelling have established themselves as multimedia and trans-media narrative strategies and are the basis of video games, graphic novels, films, and television productions, sometimes connecting a single brand within a big franchise. This trans-mediality significantly increases a brand's chances of success. For instance, *V for Vendetta* is a British graphic novel that was published in black and white between 1982 and 1985. It depicts a dystopian and post-apocalyptic future in which the United Kingdom, following a nuclear war, is dominated by a totalitarian and fascist regime. It inspired a movie (*V for Vendetta*, 2005, James McTeigue), which later became a cult. A color version of the graphic novel is now distributed in the United States.

This extraordinary success risks undermining the communicative power of dystopian and post-apocalyptic narration. In 1995, Lawrence Buell wrote that the apocalypse is perhaps "the single most powerful metaphor that contemporary environmental imagery has at its disposal" (L. Buell 1995, 285). Less than ten years later, Frederick Buell published *From Apocalypse to Way of Life: Environmental Crisis in the American Century* (2004) in which he underlines the risk of excessive recourse to the metaphor of the apocalypse, claiming that it would suffer "normalization," and accustom the public to a future when environmental disaster is the norm rather than an event to be feared. In this future, apocalyptic scenarios and desolate landscapes risk being perceived as a way of life, while representation of apocalypse is turned into "a source of excitement, not dismay, a stimulus to thrilling new adventures and a path to hitherto undreamt-of new modes of being, not an account of doom and destruction" (F. Buell 2004, 228).

It is worth wondering if post-apocalyptic and dystopian scenarios still have a warning capacity for the future. The fear of a nuclear conflagration capable of sweeping away human life seems to be less acute, but a global epidemic is always a possibility, as is a dictatorial regime, and the loss of freedom linked to the spread of communication technologies. In tandem, the general conditions of the planet have progressively deteriorated due to climate change, the accumulation of waste, and the plastic in the oceans. These phenomena have an incremental character; they do not cause sudden apocalypse on a global scale, but a progression of adverse events on a local scale and severe forms of environmental injustice. Perhaps it is better to represent them as such.

Notes

1 https://edition.cnn.com/2017/05/21/opinions/deadly-bathtub-compared-to-terrorism-opinion-geltzer-easterly/index.html

2 In the preface of the new edition, Wells writes "*When the Sleeper Wakes*, whose title I have now altered to *The Sleeper Awakes*, was first published as a book in 1899 after a serial appearance in the *Graphic* and one or two American and colonial periodicals. It is one of the most ambitious and least satisfactory of my books, and I have taken the opportunity afforded by this reprinting to make a number of excisions and alterations."

Bibliography

Booth, Robert A. (2015). Organisms and Human Bodies as Contagions in the Post-Apocalyptic State. In Gurr, Barbara (ed.), *Race, Gender, and Sexuality in Post-Apocalyptic TV and Film* (pp. 17–30). Springer.

Brereton, Pat (2005). *Hollywood Utopia: Ecology in Contemporary American Cinema*. Intellect Books.

Broderick, Mick (1993). Surviving Armageddon: Beyond the Imagination of Disaster. *Science Fiction Studies* 20(3), 362–382. www.jstor.org/stable/4240277

Buell, Frederick (2004). *From Apocalypse to Way of Life: Environmental Crisis in the American Century*. Routledge.

Buell, Lawrence (1995). *The Environmental Imagination: Thoreau, Nature Writing and the Formation of American Culture*. Princeton University Press.

Claeys, Gregory (2010). The Origins of Dystopia: Wells, Huxley and Orwell. In Claeys, Gregory (ed.), *The Cambridge Companion to Utopian Literature* (pp. 107–134). Cambridge University Press.

Claeys, Gregory (2016). *Dystopia: A Natural History*. Oxford University Press.

Collier, Patrick C. (2008). "Our silly lies": Ideological Fictions in M. Night Shyamalan's "The Village." *Journal of Narrative Theory* 38(2), 269–292. jstor.org/stable/41304887

Dini, Rachele (2019). "Resurrected from its Own Sewers": Waste, Landscape, and the Environment in JG Ballard's 1960s Climate Fiction. *ISLE: Interdisciplinary Studies in Literature and Environment*, isz003. https://doi.org/10.1093/isle/isz003

Gergan, Mabel, Smith, Sara, & Vasudevan, Pavithra (2018). Earth Beyond Repair: Race and Apocalypse in Collective Imagination. *Environment and Planning D: Society and Space* 38(1), 91–110. https://doi.org/10.1177/0263775818756079

Hillegas, Mark (1967). *The Future as Nightmare: HG Wells and the Anti-Utopians*. Oxford University Press.

Ketterer, David (1989). Margaret Atwood's "The Handmaid's Tale": A Contextual Dystopia. *Science Fiction Studies* 16(2), 209–217. [www.jstor.org/stable/4239936

Killingsworth, M. Jimmie, & Palmer, Jacqueline S. (1996). Millennial Ecology: The Apocalyptic Narrative from Silent Spring to Global Warming. In Herndl, Carl G., & Brown, Stuart C. (eds.), *Green Culture: Environmental Rhetoric in Contemporary America* (pp. 21–45). The University of Wisconsin Press.

Kunkel, Benjamin (2008). Dystopia and the End of Politics. *Dissent* 55(4), 89–98. https://doi.org/10.1353/dss.2008.0072

Lemire, Elise Virginia (2002). *"Miscegenation": Making Race in America*. University of Pennsylvania Press.

Leikam, Susanne (2017). Of Storms, Floods, and Flying Sharks: The Extreme Weather Hero in Contemporary American Culture. *RCC Perspectives* 4, 29–36. www.jstor.org/stable/e26241446

Sánchez-Escalonilla, Antonio (2010). Hollywood and the Rhetoric of Panic: The Popular Genres of Action and Fantasy in the Wake of the 9/11 Attacks. *Journal of Popular Film & Television*, 38(1), 10–20. https://doi.org/10.1080/01956050903449640

Suvin, Darko (1977). *Metamorphoses of Science Fiction*. Yale University Press.

Tower Sargent, Lyman (1976). Themes in Utopian Fiction in English Before Wells. *Science Fiction Studies* 3(3), 275–282. www.jstor.org/stable/4239043

Tuan, Yi-Fu (1979). *Landscapes of Fear*. University of Minnesota Press.

Yanarella, Ernest J. (2001). *The Cross, the Plow and the Skyline: Contemporary Science Fiction and the Ecological Imagination*. Universal-Publishers.

Zaslow, Emilie (2009). *Feminism, Inc.: Coming of Age in Girl Power Media Culture*. Springer.

5 Dystopian settings and (post)human landscapes

5.1 Settings and landscapes

"The distinction between setting and landscape, one might say, is one of pictorial economy: as long as natural space in a work is subservient to characters, events and action, as long as its function is to provide space for them, the work is not properly speaking a landscape" (Lefebvre 2011, 64). A landscape is more than just a setting; it is a complex idea, expressing the link between human beings and their context, and has a cultural and subjective dimension. It is a "view" that contains an interpretation, linked to a social and aesthetic education (Guerin 1995). It is also a way of seeing/representing through which the author frames the protagonists and their relations with the Earth and other human beings (Cosgrove 1984). A literary landscape is therefore not a simple description of the context, but a relational dimension between that context and the character (Jakob 2005). Likewise, in cinema, a landscape is more than a location or a setting. The distinction is essential, even if it does not envisage opposition as much as continuity.

Even a setting can perform several functions linked to the narrative. It can be "neutral," i.e., not offer any element, it can be "informative," or add "sympathetic" elements, or even become, in some way, a "protagonist," "that is, it may enter integrally into the action of the story" (Freeburg 1918, 151). Dystopian narratives are mostly focused on human events. The setting is just a device that emphasizes the estrangement of the narrative or provides some informative or sympathetic elements. Generally speaking, instead of offering the consolatory image of pristine nature, used in other literary and cinematographic genres to promote the diffusion of environmental instances in an elegiac-preservationist tone (Brereton 2005), dystopian settings are usually artificial and over-urbanized. Contrast is sometimes created between the oppressive setting and some images of uncontaminated and joyful nature surfacing in the protagonist's memories, or using a language rich in metaphors and similes related to natural subjects to describe life within a heavily artificial framework. Otherwise, there is a contrast between the dystopian space organized by power and the external, non-anthropized space which offers a place to escape or a last margin of freedom. The hope that there is a different reality at the end of "the road" or beyond "the border" that is not yet oppressed by totalitarian power or devastated by humanity's misconduct is a cliché, as is the final escape to the "green place."

This contrast between nature and culture is typical of the persistence of a vision marked by the dualism characteristic of the Western cultural tradition and still anthropocentric. In apocalyptic and post-apocalyptic representations, on the other hand, the setting is often a fundamental element of the narrative, to the point of becoming an "actant" or even an "intentional" landscape (Lefebvre 2011), i.e., a representation as important as the human protagonists and as significant from a diegetic point of view. The post-apocalyptic setting is often a place known to have been very crowded (a city like London or New York), now empty of people but littered with everyday objects that no longer have any use there. Alternatively, it is a space characterized by a wild regrowth of vegetation.

When nature regains the upper hand, the ability of plants and animals to take up with their apparent disorder any order imposed on them by human transformation activities highlights the precariousness of the human condition and recalls, in an ecocentric way, the Earth's ability to move forward, independent of human beings. Sometimes, the environment proves to be only partially resilient, in a compromised and impure way; for instance, in *The Drowned World* (Ballard), the rise in solar radiation causes an increase in mutations and the appearance of "freak botanical forms." In the Japanese anime, *Kaze no tani no Naushika* (*Nausicaä of the Valley of the Wind*, 1984, Miyazaki Hayao), the post-atomic forest grows luxuriant and full of color but has become a "toxic jungle." A context characterized by the slow reclamation of nature, albeit impure, is not the only kind of setting of post-apocalyptic narratives. Together with overgrown or post-natural ecology, we also have "wastelands." That is, the world may be represented as the object of total devastation, devoid of animals and plants, blasted and gray (as in the novel *The Road*, in the movie *The Book of Eli*, 2010, The Hughes Brothers, or in the video game series *Fallout*, 1997–2018). In this case, human beings are given the extraordinary power to destroy life in its totality.

These different types of setting and landscape correspond to different ways of thinking about the relationship between humankind and nature. They will be analyzed on that basis in the following paragraphs.

5.2 Green places: dreaming of "nature" in dystopian settings

An open contrast between the oppressive space occupied by human societies and "nature" is present in many dystopian classics. Jack London's *The Iron Heel* (1908) is an excellent example of this. The novel describes the advent of an authoritarian political system in the United States and a subsequent revolution that attempts to overthrow it. Although it is mainly a political dystopia, it also introduces environmental descriptions to recall the difference between the beauty of what nature offers us and the ugliness of what we do (on the first page, "the sleepy hum of bees" is contrasted with the "cries of the victims").

Sometimes, nature is gated beyond a border. In the novel *We* a gigantic Green Wall separates One State's political community from the forbidden and untamed jungle, offering a contrast between the disorder/freedom/innocence of "nature" and the imposed order of the humanized landscape. Likewise, in *Brave New World*, a portion of the planet is kept as a "Savage Reservation," because, "owing to unfavorable climatic or geological conditions, or poverty of natural resources, it has not been worth the expense of civilizing" (*Brave New World* 1932, 177). It is delimited by an electric fence. It is inhabited by "savages," that is to say by people who live in accordance with tradition and follow the rules of "nature" rather than those imposed by the World State. By contrast, the World State is fully urbanized, because the love of nature, which keeps citizens away from the factories, has been abolished by law (Rodríguez 2014).

Far from the oppressive modes of anthropic spaces, whose artificiality metaphorically alludes to the dire condition of the characters' lives, natural spaces present themselves as a shelter, whether real or just imaginary. Consequently, places of outstanding natural beauty can become ideal refuges, as occurs in *The Iron Heel*. Similarly, in *Uglies* (2005-18, Scott Westerfeld Often), those people who rebel against the Surge, the operation destined to make them all beautiful, but in reality aimed at making them less intelligent and more compliant with the rules of the totalitarian state, take refuge in "the Smoke," a space outside the city, in the midst of the wilderness. In *The Hunger Games*, Katniss finds her strength and ability to rebel against the system precisely because of her wilderness skills. The members of the resistance in *Colony*, fleeing from a desolate Los Angeles, find refuge in the pristine forests of California.

Sometimes, at the end of the story, the protagonist manages to escape. With a transition from grayish atmospheres and artificial lights, the scenery is colored by strong sunshine and the blue of the sky. Space "outside" apparently consists of a sort of "Eden" that offers salvation and freedom. In *Logan's Run* (1976, Michael Anderson), the survivors of a nuclear catastrophe live inside a sort of bio-dome, which separates them from — presumably — the contaminated external environment. Inside the dome, population growth is strictly controlled: births occur by cloning, and, at the age of 30 (though it is 21 in the novel of the same title by William Francis Nolan and George Clayton Johnson), people are eliminated through the so-called Carousel, a regeneration ceremony, whose outcome the participants are unaware of. The leading man is initially assigned to ensuring the system is properly maintained. He then discovers the reality behind the "regeneration" and flees, in the company of a young woman, triggering a series of events that lead to the breakage of the bio-dome and the discovery that full and lush vegetation still exists outside it. *Blade Runner*'s setting is a city, where the sky is perpetually dark and punctuated by gas flares (Hewitt and Graham 2015), but in the "happy ending" of the 1982 release,[1] the protagonist and his partner eventually leave it to drive towards a blue sky, mountains, and woods. In the first episode of the *Terminator* saga, the final scene shows Sarah Connor, who we know is pregnant with Michael Connor, the future hero of the fight against machines, driving away into the desert in her red car, accompanied by a German shepherd and armed with a rifle. In the distance, a storm comes, framed by a romantic landscape of desert and mountains (Brereton 2005, 200).

The dream of a "green place" as the alternative to the dystopian reality is a narrative element that now and then also appears in post-apocalyptic narratives. In *The Book of Eli*, the planet is covered with nuclear ash, and the colors are dominated by shades of gray, except at the end when Eli (a man, accompanied by a woman, Solara) arrives at Alcatraz, which is now an Edenic place where a new civilization can be born. In *Waterworld*, water covers the entire planet due to climate change but some of the characters dream about a mythical "dryland" and the protagonist eventually finds an island where it will be possible to start a new "terrestrial" life. As far as the environmental message goes, this cliché operates in a rather elegiac sense as part of the cautionary tale. It assumes or at least indicates the hope of a future way for humanity to continue. Unfortunately, this is not necessarily true, as is demonstrated by Furiosa and her group of runaways in *Mad Max Fury Road*. In her flight from the barren waterless land dominated by Immortan Joe, Furiosa heads for the place of her origins. She calls it the "Green Place" because in her memory it is green and full of trees. However, she discovers that the Green Place no longer exists; it has turned into a swampland. In this way, not only "Fury Road's narrative structure refuses to root itself in environmental nostalgia" (Yates 2017, 359), but it clearly shows us that there is only one planet, that there is no Planet B, except, perhaps, in our memory.

5.3 Dystopian borderscapes

If it is true that dystopia emphasizes the excessive order imposed by human beings on society (Kunkel 2008), borders transfer this order into space (Raffestin 2005). Therefore, in dystopian narratives, there is a multiplication of borders, whereas in the representation of post-apocalyptic/post-human/post-political landscapes they usually disappear, together with the whole process of the symbolic structuring of space.

Different motivations justify dystopian forms of segregation. The beautiful is separated from the ugly, the rich from the poor, the genetically perfect from the imperfect, the "normate" body from the mutant ("normate" refers to "the constructed identity of those who, by way of the bodily configurations and cultural capital they assume, can step into a position of authority and wield the power it grants them") (Garland Thompson 2017, xii). Boundaries can then be imposed at all scales. Cities are divided vertically to separate social classes, or into blocks to better control the population (as in the TV series *Colony*). States are fragmented, national borders become insuperable, even the Earth may be surrounded by a border, separating it from artificial, more pleasant, and less polluted worlds (as in the movie Elysium, 2013, Neill Blomkamp).

Here again, different anxieties produce different borders. A classic dystopia like the novel *When The Sleeper Wakes* reflects "the most obvious impact of nineteenth-century reality on literary imagery… the lesson of social division"

(Stableford 2010, 263) and spatializes this division in the vertical dimension. In the London of the future there will be the rich and powerful, who live up at the top, and the poor and powerless, down at the bottom. A similar "stratified urbanism" (Hewitt and Graham 2015) is also a feature of the entirely artificial space in the film *Metropolis* (1927, Fritz Lang). Here, for the first time on screen, enormous skyscrapers are presented to emphasize social divisions in a hyper-classist society where the laborers live underground and do hard physical work, while the "thinkers" live on the upper floors and are given the task of planning the existence of the others. Wells did not like the movie. In *The New York Times* (April 17, 1927) he wrote about *Metropolis*: "[It] is a city, we are told, of 'about one hundred years' hence. It is represented as being enormously high; and all the air and happiness are above, and the workers live, as the servile toilers in the blue uniform in *The Sleeper Awakes* lived, down, down, down below. Now far away in the dear old 1897 it may have been excusable to symbolize social relations in this way, but that was thirty years ago, and a lot of thinking and some experience intervene. That vertical city of the future we know now is, to put it mildly, highly improbable."

For urban developments, he was right. Still, "the use of the vertical axis to explore social divisions" remains a sort of symbolic choice even many years later, since, "in social terms, the vertical implies hierarchy" (Hewitt and Graham 2015, 930). In another movie that was destined to enter the collective imagination, *Blade Runner*, the connection between capitalism and verticalism is still visually present. And in *High Rise* (1976) J.G. Ballard uses the skyscraper to symbolize the malfunctioning of capitalist societies. In the sophisticated luxury enclave represented by the skyscraper, everything is articulated on the basis of social and economic boundaries. In the movie inspired by the novel (*High Rise*, 2015, Ben Wheatley) the contrast is delivered visually. At the top, there is the bright green hanging garden of the "creator" (the architect who designed the building), so big that even a white horse can graze there freely, while different shades of gray color the rest of the skyscraper (and the violent lives of its inhabitants).

Spatial segregation based on class is also represented in *Snowpiercer* here, the division is horizontal because the space available is a solar-powered train, which continues its run in a world made unlivable by a glaciation. The rich travel at the front and the poor at the back. In this case, too, the contrast between the two sections is pointed up by the use of color and lights, which are almost absent where the poor are forced to live, while the front of the train is bright and vivid.

Territorial segregation, also based on social class, is present in the novel *The Sea and Summer* (1987, George Turner), where those who have jobs belong to the privileged class of the Sweet, while those who don't (the Swill, nine-tenths of the population) are forced to live in huge, run-down skyscrapers (the novel, which is about climate change, was published in the United States with the title *The Drowning Towers*). In the movie *In Time* (2011, Andrew Niccol), urban inhabited spaces are divided into "zones" to separate those living in the "center" (once again, the rich and powerful) from the poor and powerless living in the peripheral regions. In the *Colony*, there is a "Green Zone," corresponding to the richest L.A. neighborhoods, where the most powerful representatives of the "collaborationists" live. In the *Hunger Games* franchise, spatial segregation is no longer urban but regional: the fictional State of Panem, located in North America, is divided into districts, ranked according to degrees of wealth, and characterized by severe class inequality and environmental injustice. In the movie *The Children of Men* (2006, Alfonso Cuarón), the division is political, and the border becomes national. Those who are on the right side of the UK border enjoy the freedom of movement, while those born on the wrong side are kept in large cages.

In addition to control and "security" issues, the border can also represent a marked distinction in terms of environmental quality. Sometimes, the whole planet is degraded, but only a few have the right to move off-planet to an artificial colony. In this regard, the border surrounds the entire planet: the poor remain on Earth, unable to leave. A similar instance of environmental injustice is advanced by the novel *Do Androids Dream of Electric Sheep?* (1968, Philip Dick), in which, following nuclear war, only the less well-off remain on Earth while others are encouraged to migrate to the off-world colonies (the topic disappears from *Blade Runner*, the movie adapted from it). The film *Elysium* highlights social contrast: Planet Earth is represented as an overpopulated shantytown, filled with garbage and pollution, while the artificial planet Elysium, where only the rich are allowed to live, replicates the style of upper-class neighborhoods. What merits attention, in this regard, is that the two spaces are not futuristic visions but extensions of realities present on the planet today, like an African slum on the one hand, and a privileged neighborhood of California on the other.

Generally speaking, the contrast between the freedom of movement of those who are privileged enough to live on the desirable side of the human territorial space and those who have the misfortune to be born on the other reflects the injustice of the present-day international system of border management. So, in dystopian narratives, the inclusion of borders and borderscapes (dell'Agnese and Amilhat Szary 2015), albeit on different scales, contributes to the representation of an over-territorialized world, and emphasizes issues of social segregation, international inequalities, and environmental injustice.

5.4 Wastelands: capitalism, consumerism, garbage

Beyond spatial inequalities and social segregation, another way to stigmatize the dangers of capitalism and the superpower of corporations is by including symbolic elements, such as waste, in the setting. This device is relatively recent, and, apart from Ballard's works (prescient again), it only became a cliché in the 2000s.

When the Sleeper Wakes is one of the first anti-capitalist dystopias. It is about a Londoner who falls asleep in 1897, wakes up in 2100, and discovers that in the meanwhile, a financial institution (the White Council) has "swallowed up nearly all the great ownership of the world," causing severe social conflict and social destitution. In the future London, Wells describes the ruins of the abandoned buildings of the suburbs but does not mention an accumulation of waste along the streets. Likewise, the film *Metropolis* also shows the contrast between the workers and the capital, with its skyscrapers and basements, but it shows no trash.

The themes of consumerism and the power of corporations are addressed in more depth in *Brave New World*. The novel is set in the year AF 632, where AF stands for "after Ford," because, given Ford's divine role in this future society, the numbering of years begins with the introduction of the assembly line. Political and economic power coexist: the World State governs the planet. People are conditioned, both prenatally and in early childhood, to enter a particular social class and to perform the duties assigned to that class, thus they offer no objection to their position in life. Individual problems are solved by the taking of a legalized drug called "soma." Sex is practiced without love, and entertainment is guaranteed by the "feelies," i.e., movie theaters equipped with special seats, with buttons that allow the audience to experience all the sensations felt by the characters on-screen. People are not supposed to read books so as not to trigger the risky process of free-thinking — nobody wants "lower-caste people wasting the Community's time over books." Children of lower classes are also conditioned to hate flowers because "primroses and landscapes [...] have one grave defect: they are gratuitous. A love of nature keeps no factories busy" (*Brave New World*, 1932).

Conversely, consumption is encouraged at all costs. One of the slogans used to push people to consume more and more is "Ending, not mending," because, in a system dominated by "God Ford," items should not be repaired, but should be thrown away so that new ones can be bought. However, there is no mention of where all these objects, ended and not mended, are supposed to go. The only reference to garbage is made in relation to the village in the "reserve." The social space organized by the World State is free of waste, thus demonstrating that in the 1930s, the excesses generated by consumerism were already a source of anxiety, but garbage was not.

Capitalism, consumerism and the more or less hidden persuasion of consumers by corporations are also at the heart of J.G. Ballard's *The Subliminal Man* (1963). The short story is set in an over-capitalized industrial system, where the dictates of corporations dominate the lives of all the citizens. Buying is compulsive, as hinted by the slogans blazed by enormous advertising signs. In this society of hyper-consumption, those who spend more get a "moral reinforcement" by the supermarkets, while cars and appliances are "traded in" for new models once every few months, shops are open 24 hours a day, factories also operate on Sundays and the working day lasts 12 hours (and may be extended to 14). Even food is artificial and disproportionate, so that a chicken for two may be "an economy twelve-pounder, the size of a turkey, with stylized legs and wings and an enormous breast, most of which would be discarded at the end of the meal" (and there are no dogs and cats to eat the leftovers either). In Ballard's short story, not just food is wasted. The only beautiful things in the urban areas are the big streets with their large junctions; but, behind the immense billboards, from which subliminal messages are issued to entice passers-by to buy more objects, it is possible to spot: "continuous junkyards filled with cars and trucks, washing machines and refrigerators, all perfectly workable but jettisoned by the economic pressure of the succeeding waves of discount models" (*The Subliminal Man* 1963, 569). The characters capable of unmasking the hidden dimension of the advertising messages are two men; the protagonist's wife is instead uncritically subjugated by it (thinking in critical terms about capitalism is simpler than getting rid of gender bias).

Concern about waste often returns in Ballard's writings (Dini 2016). It becomes an actor, and not just an extra, perhaps for the first time within a literary tradition, in one of his climate novels, *The Burning World* (republished as *The Drought*). In this, a terrible drought is threatening the world, which, "without rain, is drying up." Burnt and desolate, the landscape is dotted with rubbish. As early as the first page, in the drying up lake the "caking mudbank was speckled with pieces of paper and driftwood." In the cities, "Unwashed by the rain, the streets were covered with dust and scraps of paper, the sidewalks strewn with garbage." The scattered garbage is not only a sign of the defeat of the social system, it is the primary cause of the drought because industrial waste "discharged into the ocean basins during the previous fifty years" has formed a barrier over the water, preventing evaporation and disrupting the rainfall cycle. Although critics tend to stress that, more than an accusation aimed at "industrial modernity," the waste is an element that allows Ballard to exploit "the narrative potential of its deleterious effects" (Dini 2019, 2), the text resonates today as a powerful environmental warning.

Waste also appears in the setting of many post-apocalyptic films, from *The World, the Flesh and the Devil* (1959, Ranald MacDougall) to *28 Days Later* (2002, Danny Boyle), *The Road*, and the TV series *The Walking Dead* (2010–), in which unused objects, abandoned cars, overturned supermarket trolleys, and the everyday belongings of the non-existent humanity, are a staple. In two films from the early 2000s, garbage again plays a central role. The two films, *Idiocracy* and *Wall-E*, are similar in many ways, even if the first is a dystopia (it represents the negative evolution of current behavior), the second a post-apocalyptic story (the degeneration has gone so far that the planet is uninhabitable).

Idiocracy is a comedy film that had a minimal distribution, and a domestic opening of just $124,367,[2] but, due to its diffusion on alternative distribution channels, has since become a cult movie. Set in the United States of 2505, it depicts a country covered by heaps of waste so large as to outline new mountain ranges. It is this massive and continuous accumulation of garbage that sparks off the story. When the nth truck dumps even more trash on one of these mountains of rubbish, the excessive weight starts the "great garbage avalanche" that unearths a capsule in which Private Joe has lain since 2005. The soldier, who had been placed in hibernation as part of an experiment, was chosen because, in 2005, he was assessed as being a perfectly average human being. He was later forgotten. When he wakes up, 500 years later, he discovers he is "the smartest guy in the world." While he was asleep, the population exploded numerically but "intelligence continued to decline" because intelligent people, too busy following their careers, did not have children, while the others proliferated. So, in 2505, humanity is "incapable of solving even the most basic problems, like garbage, which had been stacked for centuries, with no plans whatsoever." Since he is now the smartest man in the world, Joe is appointed as Secretary of the Interior and therefore receives the task of solving a serious environmental problem: the land has become dry and sterile. The problem has a specific cause. Subjugated by advertising, everybody now believes that water is only for flushing toilets; they have replaced it virtually everywhere with an electrolytic beverage, which they use even to irrigate fields. Though not a botanist, Joe knows that plants require water to grow, while electrolyte drinks make them scorch, and he solves the problem (Seymour 2014).

Much more mainstream is the computer-animated film *Wall-E* (2008, Pixar), distributed by Walt Disney Studios, which has grossed over $500 million worldwide.[3]

Here a big corporation named "Buy n Large" feeds the world so opulently as to kill the vitality of the planet, covering it with waste. Human beings now live in a spaceship, where they are "coddled, fattened, and infantilized by an automated consumerist society where all labor is provided by robots and even movement of the human body is technologically mediated" (Booth 2015, 37). Rescue from this situation is brought by a small robot (a "Waste Allocation Load Lifter Earth-Class" — abbreviated to Wall-E), powered by solar energy, which finds a small plant alive on Earth, and with the help of a more advanced robot, called Eve ("Extraterrestrial Vegetation Evaluator"), succeeds in bringing humans back to the planet.

Both films send a conservative environmentalist message marked by the classic dualisms between man and woman, nature and culture. They both remain anchored to the usual gender stereotypes (the main character is a young male and the female character is only a supporting character). Moreover, they both attribute to a male protagonist the task of saving the Earth (passive and female) from an environmental disaster (Yates 2019). There is a difference, however, because in *Wall-E*, nature proves to have resilience in itself, albeit in posthuman perspective. This is testified by the presence of Wall-E's cockroach-pet, the only living thing to have survived on the planet. As Whitley (2008, 144) points out, "the cockroach is perhaps the ultimate symbol of survival through adaptation to changed environments, having existed for over 300 million years on earth, 300 times longer than human beings."

However, using waste as a metaphor and negotiating everyday life praxis, both movies are expressions of a "culture of dissent" against consumerism and corporate power. They both also make clear that "present-day patterns of consumption and wastefulness are suffocating the planet and making it uninhabitable" (Anderson 2012).

5.5 Post-human landscapes in biocentric/ecocentric perspectives: *The Last Man* and *Earth Abides*

We have seen that dystopia tends to impose excessive order on space (an excess of territorialization and artificiality). Post-apocalyptic narratives instead leave room for disorder. An Italian geographer, Aldo Sestini (1947), defined "space transformed by human beings" (now, we call it "territory") (Turco 2010) as a balance between the modifying actions of human beings (which we define as "territorialization") and the entropic processes of "nature." In dystopian narratives, over-territorialization erases nature, or locks it away outside a border. Conversely, in post-apocalyptic stories, the balance breaks up. Human action is powerful enough to erase the biosphere; or else, the forces of nature prevail, human constructions collapse, and skyscrapers are left empty and disused. Nature tends to take over and erase the signs left by humans.

The process involves all aspects of territorialization: both material ("reification"), organizational ("structuring"), and symbolic ("naming"). Rewilding, therefore, does not only involve the material transformation of buildings, animals, and vegetation; the political organization of the territory and toponymy also lose their meaning. Seen from this perspective, the representation of a post-apocalyptic landscape, marked by the return of wilderness, helps to convey a biocentric/ecocentric discourse, reducing the presumed supremacy of human beings on the planet.

Mary Shelley's novel *The Last Man* provides an early example of this. The narrative unfolds from the perspective of a young man named Lionel, who travels with friends across Europe. A mysterious plague begins to kill his comrades and the population at large while he inexplicably survives, until he finds himself alone on the planet. The novel does not describe the survivor's efforts after the apocalypse. It is about the events leading to the disappearance of the human race, and the last man's reactions towards these progressive transformations.

The admiration expressed by Lionel for "exhilarating beauty" of nature turns the scenery into a landscape, and continually reminds the reader of how beautiful nature is. "There, fanned by vernal airs, the Spirit of Beauty sprung from her cold repose; and, with winnowing wings and soft pacing feet, set a girdle of verdure around the Earth, sporting among the violets, hiding within the springing foliage of the trees, tripping lightly down the radiant streams into the sunny deep" (*The Last Man*, 1826).

Rich and colorful descriptions of the natural environment are numerous in the first part of the novel, when the young people are still "gay as summer insects" and immersed in the tranquility of the "divine forest." They continue to be even as devastation advances, and human beings begin to die like flies. Towards the end of the book, Lionel addresses his animal brothers, describing their similarities: they are all made of flesh and blood. Together with descriptions of natural elements, similes associating human characters with plants and animals are found throughout the text; in this way, humans are presented as part of the animal world, and likewise subject to the laws of nature.

Mary Shelley's "antianthropocentric" position (Moore 2017, 99) extends to all life on the planet, which will go on without any human beings, and eventually to the Earth itself, which will continue undaunted to turn, like a "green desert" (Morton 1996), impassive to human events. In this way, "the traditional anthropocentric vision of the world is replaced by the realization that nature has no need for humans" (Cameron 2012, 187). Seen from this perspective, extinction can come to the human species without the planet as a whole registering a great change: "Yes, this is the Earth; there is no change — no ruin — no rent made in her verdurous expanse; she continues to wheel round and round, with alternate night and day, through the sky, though man is not her adorner or inhabitant" (*The Last Man*, 1826).

Earth Abides (1949, George R. Stewart) is a post-apocalyptic novel that starts, more or less, where *The Last Man* ends, that is when the plague has run its course and almost all of humanity has disappeared from the planet. The story is set in California in the 1940s. The main character is a geographer who, while carrying out his observations in a secluded place, is bitten by a snake, loses consciousness and thus remains isolated in a lonely mountain cabin, unaware of what is happening around him. When he regains consciousness and goes down to the valley, he discovers that almost all of humanity has died, affected by "some new and unknown disease of unparalleled rapidity of spread." The origins of the disease are unknown, albeit three possibilities are offered: "it might have emerged from some animal reservoir of disease; it might be caused by some new microorganism, most probably a virus, produced by mutation; it might be an escape, possibly even a vindictive release, from some laboratory of bacteriological warfare. The last [is] apparently the popular idea." (*Earth Abides*, 1949).

In the first part of the novel, Ish is a "last man," who wanders alone in the desert lands of North America, from San Francisco to New York (which, with its "utter uselessness," epitomizes "the senselessness" of humans' past civilization) (Cummins Cogell 1978). Then, he becomes a "first man": he meets a woman, who becomes his wife, and together they start a new human community. In the following years, life in the new community flows, many children are born, then marry in turn. Initially, they live by scavenging in abandoned shops and opening canned foods; then, they learn to hunt and fish. As they acquire these "wilderness" skills, they lose those connected to civilization: they can no longer read and write. In this new group of hunter-gatherers (called The Tribe), Ish returns to play the role of "the last man." Not only is he the only intellectual — his descendants do not know how to spell or the meaning of the word "Civ-vil-eye-za-shun" — he is also the last to have seen it. He is the last one to have lived in a world dominated by human beings. Now, with human beings again cave-dwellers, the balance between the species is restored, and there are no more presumed rulers of the planet. In this way, the book makes explicit the biocentric/ecocentric attitude suggested by the title: humans have no privileged place in nature. If they go, "Earth abides."

As in *The Last Man*, in *Earth Abides* the apocalypse has no ascertained cause: it may have had an anthropogenic origin, but it may also be related to the "biological law of flux and reflux," since "the number of individuals in a species never remains constant but always rises and falls" (Cummins Cogell 1978). In this way, Stewart demonstrates that human beings follow the laws of nature; they are a species like others, like ants, rats, and grasshoppers: they multiply, they reach their maximum growth within a closed system and, when too numerous, they suffer an abrupt reduction in number, or even disappear. He highlights the animal nature of the human species, challenging the classical human–nature dichotomy, and employs "large-scale disaster to emphasize our embeddedness in non-human nature and to highlight the ideological trends that make this embeddedness dangerously invisible" (Otto 2012, 29).

In terms of gender and race, the text is deeply affected by the geopolitical discourse of its time. Ish is an "Adamic Hero" (Wells 2007, 473), who gives to Em, his female partner, no more than the role of "nurturer," while he cares about "the future of civilization." Moreover, he entrusts all his hopes for an intellectual revival of humankind in his male child, Joey, who can "keep the light burning through these dark times," not in his wife, least of all in his daughters. Ish is white, while Em "passes" for "white," but on the mother's side, she is not. She confesses in tears her "fault," but he takes the issue with humor because, after all, in times of apocalypse, one can no longer afford the luxury of worrying about such things. So, the book clearly does not overcome gender bias and reveals some racial anxieties ("it seems to be easier to imagine an apocalypse than it is to imagine a racially just and equitable world") (Booth 2015, 22). However, it tackles with critical awareness the human–nature relation by means of the "critique of the myth of human supremacy" (Otto 2012, 100), refracted in the descriptions of the landscape.

In *Earth Abides* the post-human world is "quiet." Noises produced by human beings in their daily activities have disappeared. The silence is interrupted only by "the chirp of a bird or the faint humming of an insect." Ish observes such changes during his travels in the first part of the novel, where descriptions are used to underline the discontinuity between the old world and the new. The return of wildness is marked by "the struggle between the native plants … moving back into the gardens, and the exotics which once had been planted there and carefully tended" (*Earth Abides*, 1949).

In the countryside, the signs of human presence, such as ruins, accumulate, while vegetation resumes its vigorous sprawl, and the land "gets to look more and more like what it had been before the white men came." In the cities, all is covered by dust, and markets are strewn with litter. Even in New York, useless and empty but for pigeons and stray dogs, "Fifth Avenue makes a beautiful corpse." In this post-human world, after the loss of human control, non-human animal species are "jostling for readjustment"; those genetically similar to human beings (such as monkeys and apes), and dogs that are either too domestic or closed in kennels die; conversely wild animals "take new freedoms." "In the times of civilization men had really felt themselves as the masters of creation. Everything had been good or bad in relation to man. So you killed rattlesnakes. But now nature had become so overwhelming that any attempt at its control was merely outside anyone's circle of thought. You lived as part of it, not as its dominating power" (*Earth Abides*, 1949).

Over the years, as the growth of shrubland continues, buildings are increasingly submerged by vegetation, which rises to cover first-floor windows, while the asphalt of the roads, concealed by grass, is no longer visible. The geopolitical effects on territorial order also disappear, making place names or boundaries meaningless. What is Arizona now, and how to define a state, Ish wonders, after he attempts to suggest to young Bob the right directions to reach New York. "'You may have some trouble in Arizona,' he went on. 'After you get to the mountains, but then…' 'What's Arry–? What is it?–Arry-*zone*-a?' Bob was asking, and it was a fair enough question. But Ish found himself stumped to answer it. What Arizona once had been — even that was a hard one. Had it been a certain amount of territory, or had it been essentially a corporate entity, an abstraction? Even so, how could he explain in a few words what a 'state' had been? Much less, how could he explain what Arizona now was? 'Oh,' he said finally, 'Arizona — that was just a name for that part over there beyond the river.' Then he had an inspiration, 'See, on the map it's this part inside the yellow line.' 'Yes,' said Bob, 'I suppose they had a fence around it?'" (*Earth Abides*, 1949).

While entropic processes take over the human reification and the naming and structuring of the land, the ongoing transformation becomes a habit. Only as an older man, Ish notices that the world is not only less noisy but has fewer colors: "something caught Ish's eye, and he reached out his hand and cried for it, suddenly, as a child might… It was a scarlet flower — a geranium, which had adapted itself to the new life and lived through these years. It was not the flower but the color, Ish realized, that had given him that sudden pang and made him cry out. There was not enough red in the world anymore. Being old, he could remember a world in which dyes and lights flamed with scarlet and vermilion. But now the world had sunk back into a quiet harmony of blues and greens and browns — and reds no longer blazed everywhere" (*Earth Abides*, 1949).

5.6 *The Drowned World* and the landscape as main character

The Drowned World is one of J.G. Ballard's "climate novels." Because of an increase in temperatures linked to solar instability and the reduction of the ionosphere, the polar ice caps have melted, and Planet Earth is now almost entirely covered by water. The story is set about 70 years after the beginning of the catastrophe. The main character, Robert Kerans, knows the world of the past only through the pictures in his old schoolbooks. Even the few people old enough remember the cities of the past just as "beleaguered citadels, hemmed in by enormous dykes and disintegrated by panic and despair, reluctant Venices to their marriage with the sea" (*The Drowned World*, 1962).

Meanwhile, humans have migrated to the polar zones from intermediate latitudes made uninhabitable by high temperatures. "The ecological balance" has reversed. Amphibians and reptilians, more adapted to the aquatic environment, are becoming more and more numerous, while the mammals are disappearing. Human beings, having their fertility reduced, have shrunk to a few million. Kerans, a marine biologist, belongs to the first generation after the apocalypse and grew up in the Polar Circle. With a group of other researchers and military personnel, he is now in the south to work at a biological mapping of the new flora, and to survey a city almost entirely submerged by water, of which he does not even know the name (halfway through the novel it turns out to be London).

As Amis (2014) writes: "Ballard gives *The Drowned World* the trappings of a conventional novel (hero, heroine, authority figure, villain), and equips it with a plot." The classic stereotypes of gender and race are present (the protagonist is a young man; the female character is inevitably beautiful, and there is even a Black man with a deferential attitude). The main character of the novel, however, is the landscape. Its importance is underlined by the emphasis given to the act of "looking," which begins right at the start of the novel: "Soon it would be too hot. Looking out from the hotel balcony, shortly after eight o'clock, Kerans watched the sun rise behind the dense groves of giant gymnosperms crowding over the roofs of the abandoned department stores four hundred yards on the east side of the lagoon" (*The Drowned World*, 1962).

Kerans is fascinated by the strange beauty of this world, where the past and the present collide. "In the early morning light a strange mournful beauty hung over the lagoon; the somber green-black fronds of the gymnosperms, intruders from the Triassic past, and the half-submerged white-faced buildings of the 20th century still reflected together in the dark mirror of the water, the two interlocking worlds apparently suspended at some junction in time" (*The Drowned World*, 1962). Although in fact, the remains of the previous civilization are sometimes trash: "Beautiful and serene from his balcony a few minutes earlier, Kerans realized that the lagoon was nothing more than a garbage-filled swamp" (*The Drowned World*, 1962).

Deterritorialization has made cities unrecognizable, or almost unrecognizable, even if human beings try to give them order again (with cartography). The landscape is thus a post-human landscape, scarred by ruins. The post-apocalyptic dimension is achieved by comparing what is going to be lost (music, elegance, the order imprinted by human activities on the land, the luxury of the hotel where Kerans lives), and what will follow. "The formalization of space and time" (Taylor 2002) is broken and the author plays with the idea of a recession of time. The ecological inversion brings back species dating from the Triassic period. Kerans, like many of his colleagues, starts dreaming about prehistoric times. For this reason, he feels part of a process that involves the entire planet and decides not to oppose it. Not only does Ballard blur the "the boundary between humankind and animal, but [...] insists that a human is and remains an animal" (Tait 2014, 36). The landscape recedes from post-human to pre-human (Rossi 1994). Human beings are not responsible for this disruption but cannot exclude themselves from it. Kerans understands he is part of a whole: "The entire planet is rapidly returning to the Mesozoic Period... In so far as we are part of the planet, a piece of the main, we too are returning..." In this way, Ballard's radical imagination "explores the deep implications of time, space, psychology, and evolutionary biology to dismantle anthropocentric narratives" (Baker 2008, 13, quoted by Tait 2014, 38).

5.7 *The Road* and the landscape as a corpse

The Road, a novel by Cormac McCarthy (2006), has received numerous prizes (including the Pulitzer Prize for Fiction in 2007) and triggered "a tremendous critical interest." In 2009, it was adapted in a movie (John Hillcoat), which was also well received by the critics, but was only given a lukewarm reception at the box-office ($27,639,579 worldwide).[4]

It is the story of an unnamed father and his son traveling in the American wasteland a few years after a major catastrophe. The duo travels south in the dim hope of finding a warmer climate, but the journey takes them through a desert of torched woodland and decaying structures, because, together with most of humanity, the catastrophe also wiped out the biosphere. Nature is no more, and everything is gray, dried, ashes. This all-encompassing devastation had occurred when the child was too young to remember.

They walk, dragging a supermarket trolley with them, where they put whatever they can gather along the way. In a world where there are no more animals or plants, only a few humans survive. Apart from gangs of thieves, all forms of organized society have been lost or have turned into dangerous covens of cannibals. Some of the survivors have descended into an abyss of degradation and turned to eating their children after feeding on their pets; others, taking control of weapons and means of transport, have organized themselves into gangs which now plunder the few survivors. Encounters with other people, therefore, along with cold and hunger, are a threat to survival. Faced with the disappearance of all human values, father and son struggle to remain "good guys," or, in one of McCarthy's metaphors, "to be the ones who carry the fire." In the end, laid low by fatigue, the father falls ill and dies. With an unexpectedly positive ending, the son joins a new family (a group of people who have maintained ties and affections, and a little dog) and resumes the journey.

In both the novel and the film, the landscape plays a primary role, and the act of looking is critical. As in *The Drowned World*, the novel opens with the main character looking at the landscape: "When it was light enough to use the binoculars he glassed the valley below. Everything paling away into the murk. The soft ash blowing in loose swirls over the blacktop. He studied what he could see. The segments of road down there among the dead trees. Looking for anything of color" (*The Road*, 2006).[5]

Scattered with desiccated remains, the scorched landscape is described by McCarthy as "so utterly defoliated and sterilized" that it represents "the greatest corpse of all" (Chabon 2007). In the movie, to make it look even bleaker, contrasts are made between the colors of the past, in the father's memory, and the gray vision of the present. Thus, although the landscape described by Stewart in *Earth Abides* is lush, and Cormac McCarthy's deserted and lifeless, the lack of color unites them (here too there is, at a certain point, something red that breaks the monotony, but it is not a geranium, it is a Coca-Cola can).

The lack of colors is not the only point in common between the two texts. In both cases, there are a father and a son (or one of the sons) who qualify as bearers of human values (which for Stewart are represented by culture, "the light," and for McCarthy by love and spirituality, "the fire"). All the other survivors are cavemen or cannibals. In *Earth Abides*, cities are meaningless concrete cathedrals. In *The Road*, the emphasis on the unnecessary accumulation of goods seems even more pronounced. Moreover, the fact that father and son move around, pushing a supermarket trolley, is almost a parody of a world of consumers, at present in ashes (Moore 2017, 235). The symbolic importance of map, also in literary terms (Rossetto 2014), comes back here too. In *The Road* father and son look at a map, but while the father recognizes signs and meanings, they escape the son: "We follow the road here along the eastern slope of the mountains. These are our roads, the black lines on the map. The state roads. Why are they the state roads? Because they used to belong to the states. What used to be called the states. But there's not any more states? No. What happened to them? I don't know exactly. That's a good question. But the roads are still there" *(The Road, 2006)*.[6]

However, the map retains a symbolic value: looking at it is a constant habit for the two protagonists, a sort of ritual that assigns a meaning to their journey. The child knows the names on it by heart, and this cheers him: they are going on, southwards, in the direction of the sea (even if the father cannot guarantee that the sea is still blue, as it appears on the map). The map falling to pieces, together with the objects of which it constitutes an evanescent representation, is itself a symbol of the transience of the "territorialization" imposed on the Earth, not only as regards reification, i.e., material transformation, but also as regards its denomination, i.e., its symbolic structuring.

Despite sharing some features, the three texts have different attitudes towards the environment. While *The Drowned World* and *Earth Abides* assign primacy to "nature," in *The Road*, the role of the human being is central.

In the latter, no mention is made about the reasons for the catastrophe. In the novel, some clues suggest that it was a nuclear explosion; in the film, some old news broadcast images hint at the same assumption (Ibarrola-Armendáriz 2009, 10). McCarthy, in an interview, stated that the event is purposely left unknown. "It could be anything — volcanic activity, or it could be nuclear war. It is not really important. The whole thing now is, what do you do?"[7] The interest is mostly focused on humans' ability to keep their spirit alive, despite everything. The role of the landscape, although central, is that of a dead world, the landscape itself is a corpse. After the apocalypse, life on the planet does not go forward (or back). It stops. The biosphere is gone. To human beings is attributed the power to annihilate life on the planet (beware, we are in the Anthropocene) (Moore 2017). It is a sin that all that beauty no longer exists (as the father's colored memories suggest in an elegiac tone, in contrast to the gray reality of the present). However, what counts is the lack of resources (and the whole journey of father and son is a search of food and shelter, of resources to survive). Thankfully, as suggested by the happy ending, some humans still know how to "carry the fire."

Notes

1 There are many versions of the film. In 1992, a *Director's Cut* was released, where the 1982 "happy ending", and Harrison Ford's voice-over narration, are eliminated (Begley 2004).
2 www.boxofficemojo.com/release/rl542868993/
3 www.boxofficemojo.com/release/rl3615065601/
4 www.boxofficemojo.com/release/rl1953269249/
5 This excerpt, taken from page 4 of the novel *The Road*, is quoted with permission from the publisher, Penguin RandomHouse, Request # 56125
6 This excerpt, taken from page 50 of the novel *The Road*, is quoted with permission from the publisher, Penguin RandomHouse, Request # 56125
7 Jurgensen J., 2009. "Hollywood's Favourite Cowboy," *The Wall Street Journal*, 20 November.

Bibliography

Amis, Martin (2014). Introduction. In Ballard, J.G. *The Drowned World*. Harper Collins.

Anderson, Christopher Todd (2012). Post-Apocalyptic Nostalgia: Wall-E, Garbage, and American Ambivalence toward Manufactured Goods. *Lit: Literature Interpretation Theory* 23(3), 267–282.

Baker, Brian (2008). The Geometry of the Space Age: J.G. Ballard's Short Fiction and Science Fiction of the 1960s. In Baxter, Jeanette (ed.), *J.G. Ballard: Contemporary Critical Perspectives* (pp.11–22). Continuum.

Begley, Varun (2004). 'Blade Runner' and the Postmodern: A Reconsideration. Literature/Film Quarterly, 32(3), 186–192.

Booth, Robert A. (2015). Organisms and Human Bodies as Contagions in the Post-Apocalyptic State. In Gurr, Barbara (ed.), *Race, Gender, and Sexuality in Post-Apocalyptic TV and Film* (pp. 17–30). Springer.

Brereton, Pat (2005). *Hollywood Utopia: Ecology in Contemporary American Cinema*. Intellect Books.

Cameron, Lauren (2012). Mary Shelley's Malthusian Objections in The Last Man. *Nineteenth-Century Literature* 67(2), 177–203.

Chabon, Michael (2007). After the Apocalypse. *The Road* by Cormac McCarthy. *The New York Review of Books*. February 15 www.nybooks.com/articles/2007/02/15/after-the-apocalypse/

Cosgrove, Dennis E. (1984). *Social Formation and Symbolic Landscape*. Croom Helm.

Cummins Cogell, Elizabeth (1978). The Middle-Landscape Myth in Science Fiction. *Science Fiction Studies* 5(2), 134–142. www.jstor.org/stable/4239175

dell'Agnese, Elena, & Amilhat Szary, Anne-Laure (2015). Borderscapes: From Border Landscapes to Border Aesthetics. *Geopolitics* 20(1), 4–13. [https://doi.org/10.1080/14650045.2015.1014284]

Dini, Rachele (2016). *Consumerism, Waste, and Re-Use in Twentieth-Century Fiction*. Palgrave Macmillan.

Dini, Rachele (2019). "Resurrected from its Own Sewers": Waste, Landscape, and the Environment in JG Ballard's 1960s Climate Fiction. *ISLE: Interdisciplinary Studies in Literature and Environment*, isz003. 165–188. https://doi.org/10.1093/isle/isz003

Freeburg, Victor Oscar (1918). *The Art of Photoplay Making*. Macmillan.

Garland Thompson, Rosemary (2017), Introduction: From Wonder to Error. A Genealogy of Freak Discourse in Modernity. In Garland Thompson, Rosemary (ed.), *Freakery: Cultural Spectacles of the Extraordinary Body* (pp. 1–19). New York University Press.

Guerin, Jean-Paul (1995). *Il grande ritorno del paesaggio*. In Calogero Muscarà (ed.), *Piani, parchi, paesaggi* (pp.121–126). Laterza.

Hewitt, Lucy, & Graham, Stephen (2015). Vertical Cities: Representations of Urban Verticality In 20th-Century Science Fiction Literature. *Urban Studies* 52(5), 923–937. [https://doi.org/10.1177/0042098014529345]

Ibarrola-Armendáriz, Aitor (2009–2010). Crises Across the Board in Cormac McCarthy's *The Road*. *Revista de Estudios Norteamericanos* 14, 81–105.

Jakob, Michael (2005). *Paesaggio e letteratura*. Leo S. Olschki.

Kunkel, Benjamin (2008). Dystopia and the End of Politics. *Dissent* 55(4), 89–98. https://doi.org/10.1353/dss.2008.0072

Lefebvre, Martin (2011). On Landscape in Narrative Cinema. *Canadian Journal of Film Studies* 20(1), 61–78. www.jstor.org/stable/24411855

Moore, Bryan L. (2017). *Ecological Literature and the Critique of Anthropocentrism*. Springer.

Morton, Timothy (1996). Shelley's Green Desert. *Studies in Romanticism* 35(3), 409–430. www.academia.edu/1077542/Shelleys_Green_Desert

Otto, Eric C. (2012). *Green Speculations: Science Fiction and Transformative Environmentalism*. Ohio State University Press.

Raffestin, Claude (2005). Confini e limiti. In dell'Agnese, Elena, & Squarcina, Enrico (eds.), *Europa. Vecchi confini e nuove frontiere* (pp. 5–16). Utet.

Rodríguez, Ángel Galdón (2014). Espacios urbanos y naturales como escenarios opuestos en la literatura distópica. *Ángulo Recto: Revista de estudios sobre la ciudad como espacio plural* 6(2), 85–100. [https://doi.org/10.5209/rev_ANRE.2014.v6.n2.47585]

Rossetto, Tania (2014). Theorizing maps with literature. *Progress in Human Geography* 38(4), 513–530. https://doi.org/10.1177/0309132513510587

Rossi, Umberto, R.M.P. (1994). Images from the Disaster Area: An Apocalyptic Reading of Urban Landscapes in Ballard's The Drowned World and Hello America. *Science Fiction Studies* 21(1), 81–97. www.jstor.org/stable/4240309

Sestini, Aldo (1947). Il paesaggio antropogeografico come forma d'equilibrio. *Bollettino della Società geografica italiana* 81, 1–8.

Seymour, Nicole (2014). Irony and Contemporary Ecocinema. In Weik von Mossner, Alexa (ed.), *Moving Environments: Affect, Emotion, Ecology, and Film* (pp. 61–78). Wilfrid Laurier University Press.

Stableford, Brian (2010). Ecology and Dystopia. In Claeys, Gregory (ed.), *The Cambridge Companion to Utopian Literature* (pp. 259–281). Cambridge University Press.

Tait, Adrian (2014). Nature Reclaims Her Own: J.G. Ballard's The Drowned World. *Australian Humanities Review* 57, 25–41. http://australianhumanitiesreview.org/2014/11/01/nature-reclaims-her-own-j-g-ballards-the-drowned-world/

Taylor, Jonathan S. (2002). The Subjectivity of the Near Future. Geographical Imaginings in the Work of J.G. Ballard. In Kitchin, Rob, & Kneale, James (eds.), *Lost in Space. Geographies of Science Fiction* (pp. 90–103). Continuum.

Turco, Angelo (2010). *Configurazioni della territorialità*. Franco Angeli.

Yates, Michelle (2017). Re-casting Nature as Feminist Space in Mad Max: Fury Road. *Science Fiction Film & Television* 10(3), 353–370. muse.jhu.edu/article/674420.

Wells, Elizabeth (2007). Earth Abides: A Return to Origins. *Extrapolation* 48(3), 472–481. https://go.gale.com/ps/anonymous?id=GALE%7CA176374797&sid=googleScholar&v=2.1&it=r&linkaccess=abs&issn=00145483&p=AONE&sw=w

Whitley, David (2008). *The idea of nature in Disney animation: From Snow White to WALL-E*. Ashgate.

Wright, Will (1977). *Sixguns and Society: A Structural Study of the Western*. University of California Press.

Yates, Michelle (2017). Re-casting Nature as Feminist Space in Mad Max: Fury Road. *Science Fiction Film & Television* 10(3), 353–370. muse.jhu.edu/article/674420.

Yates, Michelle (2019). Whiteness, Masculinity, and Environmental Nostalgia in Soylent Green and WALL-E. In Tidwell, Christy, & Barclay, Bridgitte (eds.), *Gender and Environment in Science Fiction* (pp. 167–184). Lexington Books.

6 Gulliver and beyond

Gender, race and "environmental" clichés

6.1 The "heroic male agent": white, male, young, heterosexual, and non-disabled

The protagonist of the first utopian tales is a traveler; he is a white male, in his prime and in excellent health. We could call him Gulliver. In adventure novels, the protagonist is a young (perhaps a little younger) white male, with a "normate" body (Garland Thompson 2017). We could call him Robinson. The protagonists of Julius Verne's and Wells' novels are also white males in the prime of their lives. In post-apocalyptic narrations, the hero may be a wanderer who comes to a community, solves its problems and leaves. In this way, he appeals to the old myth of "the Judeo-Christian messianic hero who battles an antichrist and his followers, liberating an oppressed community and thereby enabling social rebirth" (Broderick and Ellis 2019, 82) but also to the classical western plot structure (Wright 1977). We could call him Max. In dystopian narratives, the main character is often an active part of the system (a policeman, an officer) (*Brave New World*, 1932, Aldous Huxley; *Fahrenheit 451*, 1953, Ray Bradbury; *Soylent Green*, 1973, Richard Fleischer; *Logan's Run, Blade Runner,* 1982, Ridley Scott; *Blade Runner 2049*, 2017, Denis Villeneuve). He notices that something is wrong, he rebels, and escapes to the green place, in the company of a female partner. We could call him Deckard. Both Max and Deckard are white males and heterosexual. The "male heroic agent" is thus connoted by gender bias, which also feeds on race and age discrimination. He also respects body stereotypes, with few exceptions (Max, in the *Mad Max* franchise, transitions "from able-bodied masculinity in the first film to a limping, maddened road warrior in *The Road Warrior*") (Broderick and Ellis 2019).

In many eco-disaster/post-apocalyptic films, the protective side of the male hero is further accentuated by his role as a father (Brereton 2005; Sturgeon 2009), who "affirms the ultimate 'masculine quest' to save and protect his own flesh and blood" (Brereton 2015). So, in *Earth Abides* (1949, George R. Stewart), Ish is a father, playing Adam's role with regard to the future of humanity. *The Road* (2006, Cormac McCarthy) is the story of a father and his son. The climate change movie *par excellence*, *The Day After Tomorrow* (2004, Roland Emmerich) stars Dennis Quaid as a paleoclimatologist called Jack Hall, who, from Washington, reaches New York on foot, in the middle of a sudden glaciation, to save his son. The figure of the father, played by Tom Cruise, becomes central in Steven Spielberg's adaptation of *War of the Worlds* (2005), and in *After Earth* (2013, M. Night Shyamalan), played by Will Smith. The TV drama *The Walking Dead* (2010 –, AMC) shows a multifaceted trend in this direction. Rick Grimes, the pseudo-sheriff who leads the bunch of survivors, is a father; the narrative focuses, season after season, first on the relationship between him and his son Carl, who always wears Rick's cowboy hat as proof of his future leadership, and then with his daughter Judith, who at some point in the story inherits both the hat and the role. Even the video game *The Last of Us* (2013) has a father as the protagonist, first committed to protecting his daughter, then a girl immune to the fungal infection that has exterminated a large part of humanity.

The "green father" trope highlights the masculine protective capacity towards figures identifiable with an "authentic and primitive nature," such as representatives of tribal peoples, women, children, animals (Ingram 2000, 3). Together with "nature," they all may be victims and the object of concern and care. The heroic male agent can also play the role of the "white savior" (Hughey 2012). In this case, the encounter with a non-white supporting character (a woman, a child, a friend) emphasizes his role. In *Elysium* (2013, Neill Blomkamp), Max, the main character, is almost the only white man living on Earth; he is also the one who saves all the others. In *The Children of Men* (2006, Alfonso Cuarón), the "male heroic agent" is Theo Faron, played by Clive Owen, who is "about as stereotypical a white guy as one can get: British, confident, laconic, blue eyes, and unapologetic." His essential task in the movie is "to save Kee, the brown-skinned and unconvincingly helpless female refugee" (Hamner 2015, 1437–38), who happens to be the first woman to become pregnant after 18 years. Likewise, in the film *Snowpiercer* (2013, Bong Joon-ho), the heroic white male struggles to change the situation of the oppressed and ends up defending a child of color and his mother (however, in the TV series of the same title, 2020-on, he is instead a Black man). Concerning race, a figure similar, but opposite, to that of the white savior is that of the "magical negro" (Hughey 2012); the white savior, however, is the protagonist, while he is only a supporting figure. In addition to dealing with the problems of human "victims," the heroic male agent sometimes has the duty of saving the planet from a seemingly irreversible environmental disaster (*Idiocracy*, 2006, Mike Judge; *Wall-E*, 2008, Pixar). In this case, we are faced with what Carolyn Merchant (1996, 2003) calls an "Edenic recovery narrative." The trope "focuses on heroic male agents, traditionally white men, who function within the cinematic narrative to imagine the restoration of both civilization and Edenic nature" (Yates 2017); consequently, it tends to feminize nature and "(re)produce a binary dichotomy of passive, female nature and active, male culture" (Yates 2017, 355).

Few exceptions occur in terms of race. The protagonist is almost always white. If he is not, he cannot present himself as the new Adam, the founder of humanity, nor can he pair up with white female figures. In one of the first films focused on the theme of survival after a nuclear catastrophe (*The World, the Flesh and the Devil*, 1959, Ranald MacDougall), the main character, Ralph, is interpreted by the Afro-American actor Harry Belafonte. One of the problems of the survivors is precisely the racial difference. Ralph survives a nuclear disaster because, at the time, he was working in a mine. When he emerges from underground, he finds an abandoned New York, and begins to wander, in a scene that recalls the literary description of *Earth Abides* and is set to become a sort of "visual standard" for the genre: the streets of the city are dotted with abandoned objects, and his voice, which cries out for someone to answer, echoes among the empty skyscrapers. After wandering for a while, he meets a white woman and, while beginning to collaborate with her, rejects the hypothesis of cohabitation. A Black man and a white woman cannot be the founders of the new humanity. For this reason, when they finally meet another white man, Ralph proposes to the two whites to "start again" as a couple, entrusting them with the task of generating a new humanity, while he will continue to look for other survivors.

In 2007, Will Smith played Robert Neville in the third cinematic version of *I Am Legend* (2007, Francis Lawrence), taking the role that had previously been played by Vincent Price (1964) and Charlton Heston (1971). The roles of Will Smith and Denzel Washington, who plays Eli in *The Book of Eli* (2010, Albert Hughes, Allen Hughes), show a decline in American racial stereotypes. Again, *The Walking Dead* can serve as a parameter as the seasons unfold. "In the first three seasons, non-white characters largely play supporting, and usually disposable, roles" (Sugg 2015, 795). Afterwards, some interesting Black characters come into their own; the pseudo-sheriff, after having lost his white wife, chooses Michonne as his new partner, a powerful woman of Afro-American origin who later gets the leadership of her group.

6.2 What about the girls?

Like dystopian and post-apocalyptic narratives, sci-fi also traditionally presents a white and masculinized arena. However, because of the need to create a cognitive estrangement inherent to the genre, it sometimes attempts to challenge racialized and patriarchal gender relations (Merrick 2003). The desire to overturn the conventions typical of the "norm" in terms of race and gender relations may lead to the introduction of Black heroes. Will Smith stars as the salvific hero of humanity in sci-fi blockbusters such as *Independence Day* (1996, Emmerich), *Men in Black* (1997, Barry Sonnenfeld), *Men in Black II* (2002, Barry Sonnenfeld), and *I, Robot* (2004, Alex Proyas). Sometimes, sci-fi also introduces women-soldiers capable of taking a spaceship to safety alone, for example, Ripley (played by Sigourney Weaver) in the first episode of the *Alien* saga (1979, Ridley Scott). Alternatively, it may show planets dominated by women, who hold all the positions of power, while men have subordinate tasks, as in *Ghosts from Mars* (2001, John Carpenter). Videogames and video games-inspired franchises, like *Tomb Raider* or *Resident Evil*, may also present "Amazon-like heroines" (Merrick 2003). They are generally very tough and aggressive, and sexy and commanding enough to be able to oscillate between transgression of traditional gender roles and sexual fetishization (Lara Croft, from *Tomb Raider*, has been visually associated with a "dominatrix") (Herbst 2004).

In male-centered dystopian and post-apocalyptic narratives, the "protagonists are men, women's concerns are left unmentioned, and women characters are either trivial or vague" (Tan 2017, 32). Women are generally supporting characters, trophies to conquer or victims to protect; they are "acted upon," and their presence "is solely for the feelings and emotions [...] [they] can incite in the male" (Broderick and Ellis 2019, 55). Sometimes, they accompany the leading male characters on their adventure; in that case, they risk becoming "action chicks" (Inness 2004), sharing the "hegemonically masculine behavior" of their partners (Leikam 2017, 33), such as Carol and Michonne in the first seasons of *The Walking Dead* and Nova in *Sharknado* (2013, Anthony Ferrante). Even when they assume a warrior role, women usually lack real leadership skills and make fatal mistakes (again, Carol in *The Walking Dead*). At the most, in post-apocalyptic action movies, they play the role of an antagonist, like Aunty Entity in *Mad Max Beyond Thunderdome* (1985, George Miller and George Ogilvie). If so, they may be not humans, but androids with a feminine appearance, as in *Blade Runner 2049*: here the main character is surrounded by female figures, mostly clones, AI or replicants obeying orders of the real (male) adversary.

Since *Metropolis* (1927, Fritz Lang), feminized androids, cyborgs, and artificial intelligence are often present. They may be antagonists but also sexual partners, or, at least, sexualized partners and objects of the male protagonist's desire: for example, *Blade Runner*, with Rachael the young and beautiful replicant; *Her* (2012, Spike Jonze), with Samantha, a virtual voice who knows the desires of the main character so well as to make him fall in love with her; and *Blade Runner 2049*, with the lovely holographic figure of Joi, acting as a partner of K, the protagonist, who is himself a replicant. Being artificial makes all these female figures passive, generally obedient, and ready to indulge, verbally, or even physically.

As remarked by Christy Tidwell (2019, 22), "these films are [...] obviously ripe for discussion of gender representation. They also — less obviously — raise significant questions regarding representations of the environment, the connections between gender and environment, and the place of technology in this nexus." The critique, from the point of gender construction, is explicit: *Her* has the ultra-sexy voice of Scarlett Johansson, while Joi is beautiful and always available, a "virtual projection girlfriend" that "becomes a more dependable, more attractive decoration that satisfies the expectation of male audiences" (Tang 2019, 15). Less explicit is the connection with nature. In *Her*, nature is only evoked, via a small and incredibly sad urban park, and by some images on the walls, but it is, apparently, under control; therefore, it is, like femininity, not a problem. What is at stake is technology (and the male protagonist's ability to control it). In *Blade Runner 2049*, there are more scenes shot outdoors, in which the settings show a world ravaged by climate change. Still, the sad state of the environment appears to be a given, a taken-for-granted aspect of the future, not an issue (it is "a way of life," F. Buell 2003). If trees are dead and rural landscapes devastated, it matters little; at stake is the replicants' ability to reproduce. Again, the central theme is control over technology, not the environmental disaster (Hamblin and O'Connell 2020).

Not all dystopian and post-apocalyptic texts are so androcentric, in fact, there is a growing number of notable exceptions. An essential input to the discussion on gender issues, and the women–nature connection in dystopian narratives, is provided in Margaret Atwood's *The Handmaid's Tale* (1985), highly celebrated for its ecofeminist approach. The novel was first adapted into a film (1990, Volker Schlöndorff), into an opera (2000, Poul Ruders), and then into a television series (2017–) that won the 2017 Emmy Award for Outstanding Series. A first-person narrative, the novel is set in the future totalitarian State of Gilead, located where New England is now. Gilead is a kind of militarized theocracy, whose regime exercises strict control over the life of its population in all aspects, including the most intimate. Pollution and radiation have heavily affected human fertility so the few fertile women are offered as "handmaids" to regime members (the so-called Commanders), with whom they must bear children. To this end, Commanders rape them each month in a particular "Ceremony" also attended by their wives. While men are all dressed in black, the colors of women's clothes distinguish their wearer's duties: the wives of the Commanders are dressed in blue; the handmaids in red; the Marthas, responsible for household services, in "dull green."

The protagonist, a handmaid, is called Offred because she belongs to a Commander called Fred, not because that is her real name. In this regime of total patriarchy, where women are not allowed to write or read, "flowers are still allowed" (as Offred says at the beginning of the story). Flowers often return in the narrative; there are paper flowers, flowers painted on the walls, flowers in the garden of Serena, the Commander's wife, images of flowers that can be envisaged in the bodies of people hanged for crimes against the State (whose red reminds Offred of the tulips in Serena's garden). "Flowers, in fact, pervade the novel, signifying resistance to Gilead's agenda but also providing a backdrop for many of the novel's most explicit scenes of sexual exploitation" (Hooker 2006, 287). They thus serve as metaphors for the female body. With this continuous pairing, the novel highlights the role of the woman as a mere "resource," relevant only in relation to her body, as long as she has "viable ovaries" (in that case, she is essential and scarce, says Offred, as a real "national resource") or she can provide services as a "Martha." From an ecofeminist perspective (Kapoor 2018), the novel associates the passive roles of woman and nature in the patriarchal system to indicate that ecological imbalance and gender oppression go hand in hand.

A similar perspective also characterizes the sequel, titled *The Testaments* (2019). Here, the association between the female body and nature is made even more explicit. The idea that the girls are "precious flowers" is reiterated in the novel in different contexts, to emphasize how they cannot rebel against their fate: "We were custodians of an invaluable treasure that existed, unseen, inside us; we were precious flowers that had to be kept safely inside glass houses," and later in the book: "we were precious flowers, and who ever heard of a rebellious flower? [...] No rebellious flowers here!" (*The Testaments*, 2019).[1]

Atwood's texts primarily highlight the oppressive regime to which many women are subjected, using nature as a metaphor. More recent texts highlight women's ability to rebel and assume control, including in the relationship between human beings and the environment. In *Mad Max Fury Road* (2015, George Miller), Furiosa, the female lead, is a central figure alongside the character in the title. She is a senior member of Immortan Joe's regime, a power system in which the dictatorial leader exerts strict control over access to water in an extremely dry setting called the Citadel. Furiosa rebels and runs away, taking the "breeders," the old dictator's young wives, with her. She tries to lead her group of fugitives to the place of her childhood, which she remembers as the Green Place. However, she discovers that it is no longer green and has become as bad for human life as the Citadel (where at least there is a water source, even if Immortan Joe controls it). She returns and conquers the city, giving back the water, and thus life, to the people and the land.

Furiosa is a person with disabilty; she is not the only one with a body that is not "normate" in the movie, as Immortan Joe's army, his people, and his son, are all persons with disabilities. Many of them live thanks to machines (or continuous transfusions), and their damaged bodies are somehow emblematic of the Citadel's desolate setting. Furiosa is also a cyborg, tapping "the liberatory potential of cyborg imagery and identity for feminism" (Soles 2019). Immortan Joe is a cyborg, too, because he survives thanks to an elaborate suit of armor whose forms enhance his masculinity. Unlike Immortan Joe, Furiosa can live without a prosthesis, and, when she takes power, she does so just as a woman with a disability. In this way, she not only claims her own ability to act as a "powerful and agentive" person (Yates 2017), but also that of the team of women she leads, who are capable of acting together and jointly defeating patriarchy (Broderick and Ellis 2019). She also links women and nature, "not to serve capitalist patriarchy, but rather [...] to re-cast nature as feminist space" (Yates 2017, 369).

However, beyond the ecofeminist message in the film, some elements deserve further reflection. In the former Green Place, for instance, the environment is too degraded for human beings to tolerate, but for a vast flock of crows it is perfectly acceptable. In this way, the environmental problem risks being seen only through anthropocentric eyes. Besides, as remarked by Carter Soles (2019), in the film great use is made of fossil-fueled transportation, which in some way provides an "implicit endorsement of a laissez-faire petroleum-based economy."

Several other strong and rebellious heroines populate Young Adult Fiction and its film adaptations. The most famous are Katniss in the *Hunger Games* trilogy (2008-10, Suzanne Collins), Tally in *Uglies* (2005-07, Scott Westerfeld), and Tris in the *Divergent* series (*Divergent*, *Insurgent*, *Allegiant*, and *Ascendant*) (2011-13, Veronica Roth). In all these cases, the heroine, besides being rebellious, is also very young; this not only places her in opposition to the system (whose structural flaws in terms of class and gender relations are invariably highlighted), but also to adult generations. In some cases, these heroines bring an environmentalist discourse into play, even if they often do so only by placing the action in a degraded and over-urbanized future (Rodríguez 2017) marked by the over-exploitation of resources, in a conservationist perspective or using the traditional cliché of nature as a refuge and place of peace as opposed to the city (McDonough and Wagner 2014, 158).

In the film version, of these sagas the rebel heroine is always beautiful, respecting Hollywood clichés. She is almost always white, so much so that the failure to address the race issue has been widely raised (Couzelis, 2013; Dubrofsky and Ryalls 2014; Burke and Kelly 2015).

As with the issues of fatherhood and race, *The Walking Dead*, an audience-friendly television series, shows an interesting evolution. Season after season, many female leaders emerge and many female villains too (like Alpha, the Whisperers' leader). Even the legacy of the former leader, symbolized by the cowboy hat, has passed from the son Carl to the daughter Judith, a pre-adolescent (white) girl who exerts a great influence on the other community members.

6.3 Indigenous and settlers: "invasion fiction" and the apocalypse as historical experience

H.G. Wells' most famous work, *The War of the Worlds* (1897), is generally considered one of the foundational texts of sci-fi as a literary genre. The novel is an example of "invasion fiction." It narrates the cruel fate of Earthlings after the planet has been invaded by Martians, who are superior to them in both intelligence and technology. The novel can also be defined as a post-apocalyptic dystopia because, for the Earthlings, an event occurs that brings about a change of state (a catastrophe), following which their living conditions worsen appallingly.

Invasion fiction has become a staple of later sci-fi works, usually attributing a metaphorical sense to the trespasser. There have been countless invasions of aliens, body snatchers, giant ants, and huge apes coming out of the sea. The alien invasion trope has often been used to highlight the superiority of humans (or, in a more targeted way, Americans) over the enemy threat (Roberts 2000). The various adaptations of Wells' text aptly illustrate this point. The radio version, produced by Orson Welles in 1938, expressed the fear of a war against the Nazis; the first film adaptation, directed in 1953 by Byron Askins, reflected nuclear anxieties. The second cinematic version, directed by Steven Spielberg in 2005, turned the "plot of invasion" into the "alien terrorist" metaphor, and was explicitly marketed as a post-9/11 movie (Frank 2011).

Wells' novel, however, is much more imaginative than the later adaptations. It inverts the viewpoint, proposing to a people of invaders (the British) the prospect of being invaded and then colonized. In the preamble to the novel, Wells writes, with an admirable attack on the dominant anthropocentric vision of the late 19th century: "We men, the creatures who inhabit this earth, must be to them [the Martians] at least as alien and lowly as are the monkeys and lemurs to us." And then he adds, underlining the parallelism between British and Martian colonialism: "And before we judge of them [the Martians] too harshly, we must remember what ruthless and utter destruction our own species has wrought, not only upon animals, such as the vanished bison and dodo, but upon its own inferior races. The Tasmanians, in spite of their human likeness, were entirely swept out of existence in a war of extermination waged by European immigrants, in the space of fifty years. Are we such apostles of mercy as to complain if the Martians warred in the same spirit?" (*The War of the Worlds*, 1897).

The War of the Worlds can, therefore, be considered an allegory of British colonialism (Rieder 2012), which refers to the British invasion of Tasmania and creates empathy with those who suffered under colonialism. As remarked by Berlatsky (2014), the book suggests that the survival of the fittest is natural and unstoppable, thus introducing a Darwinian perspective to encounters between species. It also shows the transience of what is generally assumed to be permanent, such as "our" civilization, and how the "ruin of one civilization would pile up on the ruin of another" (Moore 2017, 216). The same idea of the "Course of Empire" is picked up in *Earth Abides* when Ish explains: "The world of those Indians passed away… And now our world that followed theirs has passed too."

In the opinion of some authors, all post-apocalyptic literature should be read in this way. The central role attributed to Americans in it is what Hsu and Yazel (2019, 347) call "structural appropriation": "a process in which the world-threatening structural violence that has already been experienced by colonized and postcolonial populations is projected onto American (and predominantly white) characters and readers." As remarked by Gergan, Smith, and Vasudevan (2018), with the tropes of apocalyptic narratives, "Hollywood enacts a darkly ironic reversal of the history of genocide and pestilence in the Americas. This apocalypse, which already happened, is temporally displaced into the present or future when Hollywood imagines global tragedy. Brad Pitt facing off against zombies in *World War Z* (2013, Marc Forster) folds time to give us a white protagonist fighting extinction on a continent that already witnessed a vicious effort by settler colonists to eliminate native populations."

The indigenous peoples of North America, who were probably more than 20 million at the time of Columbus' landing, and who in the twenty-first century number less than three million in the United States, have survived an apocalyptic event. "For indigenous peoples of the United States and globally, Europeans and their invasive descendants are the walking deaths they must continually evade, accommodate or fight against" (Gurr 2015, 7).

To take this point of view, the work of indigenous writers must be considered. In the collection of short stories *The Lone Ranger and Tonto Fistfight in Heaven* (1993), Sherman Alexie makes a specific reference to the cancellation of the native population following the arrival of white settlers. In particular, in the short story "A Drug Called Tradition," one of the characters says: "They're all gone, my tribe is gone. Those blankets they gave us, infected with smallpox, have killed us. I'm the last, the very last, and I'm sick, too. So very sick. Hot. My fever burning so hot" (*The Lone Ranger and Tonto Fistfight in Heaven*, 1993).[2]

In this way, Alexie refers obliquely to the smallpox epidemic of 1862 caused by the landing of a sick passenger at Victoria on Vancouver Island (Fraser 2015). This epidemic in 1862 was not a global catastrophe, but it was an apocalyptic event on a local scale. In addition to the events directly related to the invasion, other phenomena that are part of the Western dystopian and post-apocalyptic imaginary have already occurred to indigenous peoples as a result of various forms of colonialism: ecosystem collapse and loss of biodiversity, destruction of the traditional economic system, population resettlement and cultural decay (Whyte 2018, 226). For this reason, post-apocalyptic and indigenous sci-fi literature tends to convey a sense of ongoing rather than imminent crisis (Scott 2016).

In this regard, it is worth taking into account *Moon of the Crusted Snow* (2018), a post-apocalyptic novel by the Anishinaabe writer Waubgeshig Rice. At first, the novel appears as a compilation of post-apocalyptic clichés. An unspecified catastrophic event blocks communications and supplies, resulting in a breakdown of urban order and some people consider turning to cannibalism. The lead character is a young father named Evan who must protect his wife Nicole, and their children Maiingan and Nanghohns. However, there is an element of difference: the apocalyptic event does not occur in the First Nation Reserve where the novel is set, but in an unspecified elsewhere. In the Reserve, the electricity goes out, computers and mobile phones no longer work, and the supply truck does not arrive, but nothing happens. An echo of the post-apocalyptic chaos is heard in the Reserve's isolation only through the experiences of two students who have fled the city. Then, a non-native called Scott arrives. Armed with weapons and liquor, he sows disorder and division. Post-apocalyptic clichés are used to inform readers that the current "apocalyptic imagination" typical of mainstream popular culture pertains to the narrative of "the white man's system," not to the protagonists of this story. They have already lived through their apocalypse and somehow overcome it. As the village's oldest person, old Allen, says: "Apocalypse! What a silly word. I can tell you there's no word like that in Ojibwe. Well, I never heard a word like that from my elders anyway… The world isn't ending … Our world isn't ending. It already ended. It ended when the Zhaagnaash came into our original home down south on that bay and took it from us. That was our world" (*Moon of the Crusted Snow*, 2018).[3]

The ancestors of the band "were displaced from their original homeland in the South" and relocated to the north of Canada, in a harsh environment where winters are freezing, because "the white people who forced them here had never intended them to survive." They survived, nonetheless, but have seen their own culture progressively erased, their traditional ways of being almost obliterated, and now they live, depending on the outside world for energy, food, and communications. Those who still know how to hunt are few, and even the traditional language has been almost lost to English, while grandparents teach a few words to the children (the only ones who have names linked to the native tradition). Evan can still hunt, and he has a cabin in the forest. Getting out of the apocalypse perhaps means leaving the modern houses, whose appliances are of no use without energy, and instead taking refuge "in the heart of Anishinaabe territory" to recover the traditions of their own people.

Notes

1 These excerpts, taken from pages 36 and 66 of the novel *The Testaments*, are quoted with permission from the publisher, Penguin RandomHouse, Request # 56132
2 These excerpts, from *The Lone Ranger and Tonto Fistfight In Heaven*, copyright © 1993, 2005 by Sherman Alexie, are quoted with permission from the publisher GROVE/ATLANTIC, INC.
3 This excerpt, from the novel *Moon of the Crusted Snow*, Waubgeshig Rice, published by ECW Press Ltd., 2018, 9781770414006, is quoted with permission from the publisher ECW Press Ltd.

Bibliography

Berlatsky, Noah (2014). Why Sci-Fi Keeps Imagining the Subjugation of White People. *The Atlantic,* April 25. www.theatlantic.com/entertainment/archive/2014/04/why-sci-fi-keeps-imagining-the-enslavement-of-white-people/361173/

Brereton, Pat (2005). *Hollywood Utopia: Ecology in Contemporary American Cinema.* Intellect Books.

Brereton, Pat (2015). *Environmental Ethics and Film.* Routledge.

Broderick, Mick, & Ellis, Katie (2019). *Trauma and Disability in Mad Max: Beyond the Road Warrior's Fury.* Springer.

Buell, Frederick (2003). *From Apocalypse to Way of Life: Environmental Crisis in the American Century.* Routledge.

Burke, Mary C., & Kelly, Maura (2015). The Visibility and Invisibility of Class, Race, Gender, and Sexuality in The Hunger Games. In Gurr, Barbara (ed.), *Race, Gender, and Sexuality in Post-Apocalyptic TV and Film* (pp. 59–70). Springer.

Couzelis, Mary J. (2013). The Future Is Pale: Race in Contemporary Young Adult Dystopian Novels. In Basu, Balaka, Broad, Katherine R., & Hintz, Carrie (eds.), *Contemporary Dystopian Fiction for Young Adults: Brave New Teenagers* (pp. 131–144). Routledge.

Dubrofsky, Rachel E., & Ryalls, Emily D. (2014). The Hunger Games: Performing Not-Performing to Authenticate Femininity and Whiteness. *Critical Studies in Media Communication* 31(5), 395–409.
[https://doi.org/10.1080/15295036.2013.874038]

Frank, Michael C. (2011). Alien Terrorists: Public Discourse on 9/11 and the American Science Fiction Film. In Hammond, Phil (ed.), *Screens of Terror: Representations of War and Terrorism in Film and Television since 9/11* (pp. 149–169). Arima Pub.

Fraser, Gordon (2015). Troubling the Cold War Logic of Annihilation: Apocalyptic Temporalities in Sherman Alexie's The Lone Ranger and Tonto Fistfight in Heaven. *PMLA* 130(3), 599–614. https://doi.org/10.1632/pmla.2015.130.3.599

Garland Thompson, Rosemary (2017). Introduction: From Wonder to Error. A Genealogy of Freak Discourse in Modernity. In Garland Thompson, Rosemary (ed.), *Freakery: Cultural Spectacles of the Extraordinary Body* (pp. 1–19). New York University Press.

Gergan, Mabel, Smith, Sara, & Vasudevan, Pavithra (2018). Earth Beyond Repair: Race and Apocalypse in Collective Imagination. *Environment and Planning D: Society and Space* 38(1), 91–110, https://doi.org/10.1177/0263775818756079

Gurr, Barbara (ed.) (2015). *Race, Gender, and Sexuality in Post-Apocalyptic TV and Film.* Springer,

Hamblin, Sarah, & O'Connell, Hugh C. (2020). Blade Runner 2049's Incongruous Couplings: Living and Dying in the Anthropocene. *Science Fiction Film & Television* 13(1), 37–58. www.muse.jhu.edu/article/750562.

Hamner, M. Gail (2015). Sensing Religion in Alfonso Cuarón's "Children of Men." *Religions* 6(4), 1433–1456. https://doi.org/10.3390/rel6041433

Herbst, Claudia (2004). Lara's Lethal and Loaded Mission: Transposing Reproduction and Destruction. In Inness, Sherrie (ed.), *Action Chicks: New Images of Tough Women in Popular Culture* (pp. 21–45). Palgrave Macmillan.

Hooker, Deborah (2006). (Fl)orality, Gender, and the Environmental Ethos of Atwood's "The Handmaid's Tale." *Twentieth Century Literature* 52(3), 275–305. https://doi.org/10.1215/0041462X-2006–4001

Hsu, Hsuan, &Yazell, Bryan (2019). Post-Apocalyptic Geographies and Structural Appropriation. In Morgan, Nina, Hornung, Alfred, & Tatsumi, Takayuki (eds.), *Routledge Companion to Transnational American Studies* (pp. 347–356). Routledge.

Hughey, Matthew W. (2012). Racializing Redemption, Reproducing Racism: The Odyssey of Magical Negroes And White Saviors. *Sociology Compass* 6(9), 751–767. https://doi.org/10.1111/j.1751-9020.2012.00486.x

Ingram, David (2000). *Green Screen: Environmentalism and Hollywood Cinema.* University of Exeter Press.

Inness, Sherrie (2004). *Action Chicks: New Images of Tough Women in Popular Culture.* Springer.

Kapoor, Ragini (2018). An Ecofeminist Study of Margaret Atwood's The Handmaid's Tale. In Giri, Dipak (ed.), *Postcolonial English Literature: Theory and Practice* (pp. 104–114). Authorpress.

Leikam, Susanne (2017). Of Storms, Floods, and Flying Sharks: The Extreme Weather Hero in Contemporary American Culture. *RCC Perspectives* 4, 29–36. www.jstor.org/stable/e26241446

McDonough, Megan, & Wagner, Katherine A. (2014). Rebellious Natures: The Role of Nature in Young Adult Dystopian Female Protagonists' Awakenings and Agency. In Nelson, Claudia (ed.), *Female Rebellion in Young Adult Dystopian Fiction* (pp. 157–170). Ashgate.

Merchant, Carolyn (1996). Reinventing Eden: Western Culture as a Recovery Narrative. In Cronon, William (ed.), *Uncommon Ground: Rethinking the Human Place in Nature* (pp.132–159). WW Norton & Company.

Merchant, Carolyn (2003). *Reinventing Eden: The Fate of Nature in Western Culture.* Routledge.

Merrick, Helen (2003). Gender in Science Fiction. In James, Edward, & Mendlesohn, Farah (eds.), *The Cambridge Companion to Science Fiction* (pp. 241–252). Cambridge University Press.

Moore, Bryan L. (2017). *Ecological literature and the critique of anthropocentrism.* Springer.

Rieder, John (2012). *Colonialism and the Emergence of Science Fiction.* Wesleyan University Press.

Roberts, Adam (2000). *Science Fiction.* Routledge.

Rodríguez, Ángel Galdón (2017). Recurrent Dystopian Themes in Scott Westerfeld's Novel 'Uglies'. *Angloamericanae Journal* 1(1), 61–84. https://aaj.ielas.org/index.php/Angloamericanae/article/view/22

Scott, Conrad (2016). (Indigenous) Place and Time as Formal Strategy: Healing Immanent Crisis in the Dystopias of Eden Robinson and Richard Van Camp. *Extrapolation* 57(1–2), 73–93. https://doi.org/10.3828/extr.2016.6

Soles, Carter (2019). Mad Max. Beyond Petroleum? In Tidwell, Christy, & Barclay, Bridgitte (eds.), *Gender and Environment in Science Fiction* (pp. 185–201). Lexington Books.

Sturgeon, Noël (2009). *Environmentalism in Popular Culture: Gender, Race, Sexuality, and the Politics of the Natural.* University of Arizona Press.

Sugg, Katherine (2015). The Walking Dead: Late Liberalism and Masculine Subjection in Apocalypse Fictions. *Journal of American Studies* 49(4), 793–811. https://doi.org/10.1017/S0021875815001723

Tan, Cenk (2017). Rebellious Women in Men's Dystopia: Katniss and Furiosa. *Pamukkale Üniversitesi Sosyal Bilimler Enstitüsü Dergisi* 26, 32–46. [https://doi.org/10.5505/pausbed.2017.22599]

Tang, Wenjia (2019). A Research on the Change of Female Characters Portrayed in Science Fiction Films and Its Social Influences. *International Journal of Culture and History* 5(4), 14–17. [https://doi.org/10.18178/ijch.2019.5.4.140]

Tidwell, Christy (2019). Either You're Mine or You're Not Mine: Controlling Gender, Nature, and Technology in *Her* and *Ex Machina*. In Tidwell, Christy, Barclay, Bridgitte (eds.), *Gender and Environment in Science Fiction* (pp. 21–44). Lexington Books.

Whyte, Kyle Powys (2018). Indigenous Science (Fiction) for the Anthropocene: Ancestral Dystopias and Fantasies of Climate Change Crises. *Environment and Planning E: Nature and Space* 1(1–2), 224–242.

Wright, Will (1977). *Sixguns and Society: A Structural Study of the Western.* University of California Press.

Yates, Michelle (2017). Re-casting Nature as Feminist Space in Mad Max: Fury Road. *Science Fiction Film & Television* 10(3), 353–370. [www.muse.jhu.edu/article/674420.

3

Posthuman worlds

7 Post-human/transhuman/posthuman

7.1 (Post)human wor(l)ds?

"Posthuman is a seductive term, being both nebulous and popular" (Gray 2017, 148). Posthumanism is a seductive term too. Both are often used as umbrella terms (Ferrando 2013) to cover a wide range of meanings in philosophical studies and the fields of science, literary criticism, cultural studies, and more. The nebulosity manifests itself through the lack of a univocal spelling. "Posthuman" and "posthumanism" as compound words prevail in research; "post-human" is more widespread in everyday language. Confusion increases since, though they are often used synonymously, the two expressions may mean very different things.

With regard to the word "posthuman," Cary Wolfe writes: "The first time I used it (hyphenated, no less) was in an essay written in 1995, called 'In Search of Post-Humanist Theory'" (2010, xii). The phrase in brackets ("hyphenated, no less") is critical. The *hyphenated* form suggests a separation, and the prefix "post-" implies a temporality (subsequent to). The compound word, on the other hand, suggests a concurrence and a merging. Thus "post-human" is something that comes after the human, a "break with humanism" (Ranisch and Sorgner 2015), while "posthuman" suggests the need to overcome the binary distinction between the human and the "other," whatever the other may be (and then, metaphorically, to overcome all boundaries between binary oppositions). Another pair of words, "transhuman" and "transhumanism," also contributes to confusion, because, even if they are not synonymous with the first two (or rather, they are, in a scientific sense, very different "beasts") (Lemmens 2015), they are at times used in this sense.

As for "post-human," namely "that which comes after the human," it should be noted that, following Carl Linnaeus' nomenclature, the contemporary human being belongs to the species *Homo sapiens*. *Homo sapiens* is the only human species surviving out of the more than 20 species of "archaic humans" that existed in prehistory (the most famous of which are probably the Neanderthals and Denisovans). From this perspective, speaking about "post-humans" may refer to "post-human speciation, as the next iteration of the human, after H. sapiens" (Gray 2017, 139). Therefore, the term "post-human" has to do with evolution, or preferably with the possible future evolution of *Homo sapiens* in other species. For this, Gray (2017) suggests replacing the term "post-human" with the more clear-cut term "post-sapiens" (even if it must be added that the same Linnaeus definition may be criticized as being gendered and speciesist, from a posthuman point of view) (Ferrando 2015).

Otherwise, "transhumanism is the project of modifying the human species via any kind of emerging science, including genetic engineering, digital technology, and bioengineering" (LaGrandeur 2015, 112). It advocates the technological enhancement of humans, or, better, the transition of humans "into other creatures that live longer (someday perhaps forever), can live in space, and generally are engineered to be significantly different from homo sapiens" (Gray 2017, 140).

Both issues relate to the evolution of the human being (post-human), or its technological improvement (transhuman). However, transhumanism results may be referred to as "the posthuman" (LaGrandeur 2015; Ranisch and Sorgner 2015), while posthumanism goes further than that. It considers the posthuman product of transhumanism (the cyborg) as a metaphor (Haraway 1985), and "the increasing technologization of the human as a reason to question traditional humanist conceptions of the human being" (Lemmens 2015, 432). Posthumanism calls into question humanism as a "Eurocentric paradigm" (Braidotti 2013, 13) and the "allegedly abstract ideal of Man as a symbol of classical Humanity," not only a male of the species but also "white, European, handsome and able-bodied" (Braidotti 2013, 24). Moreover, posthumanism underlines that there is never a neutral subject, even if "traditionally, scientific observations have been elaborated from a specific standpoint… white, Western, economically privileged, heterosexual, and male" (Ferrando 2015, 218).

Defined as "a hybrid of machine and organism, a creature of social reality as well as a creature of fiction" (Haraway 1985; 2016, 5), the "cyborg" emphasizes the necessity of doing away with three "crucial boundaries": those between nonhuman, human and animal, organism and machine, and physical and nonphysical. The cyborg is "a creature of a postgender world" (Haraway 1985, 2016, 8), a metaphor of the need and ability to overcome the binary oppositions typical of Western society, such as male/female, nature/culture, human/animal. Posthumanism represents an attempt to avoid the affirmation of all categorical dualities (because "sexism is not separated from speciesism, biocentrism and so on") (Ferrando 2014, 170). As such, posthumanism may be conceived as a "theoretical frame, as well as an empirical one, which can apply to any field of enquiry, starting from our location as a species, to the individual gaze" (Ferrando 2014, 168).

"Any form of discrimination is a potential carrier for any other forms of discrimination, and it is related to all forms of discrimination" (Ferrando 2014, 170). In this perspective, posthumanism helps us to revisit also the binary geographies produced by the modern geopolitical vision (Agnew 2003). In geopolitical terms, the posthuman approach identifies (and overcomes) processes of construction of the self and the Other, and consequently, the dynamics of the identification of social and racial categories in a hierarchical sense and the identification of the enemy. If differences are not to be essentialized, spatial identities too are relational and not fixed; consequently, being a friend or an enemy does not depend on where one lives, but on the practices of interactions. Even boundaries blur, while "State territory" is no more to be conceived as a rigid container of sovereignty, limited by equally rigid borders, but as a bounded portion of relational space (dell'Agnese 2013, 122).

Therefore, if these words were assigned meanings to help make distinctions of a theoretical nature, it would be possible to say that the post-human comes after the human (for the better, or maybe for the worse); transhumanism foresees an improvement of the human being's potentialities, through technology; posthumanism starts from the product of the transhumanism (the cyborg) and metaphorically uses it, to emphasize the need to overcome all binary categories of differentiation. The image of the cyborg as a being capable of straddling many different binary systems gives it a considerable "metaphorical resonance" (Surkan 2004, 114) not only in critical studies but also in popular culture.

Sci-fi likes to imagine how human beings can evolve, for better or for worse, so much so that it is one of the main narrative strategies to visualize the post-human and transhuman (Gray 2017). Dystopian and post-apocalyptic narratives like to reason about the possible modifications to humankind. The factors that can trigger change are many and different. They vary from evolutionary factors, modifying the species in a post-sapiens sense, to mysterious epidemics and the transformation of humans into undead monsters devoid of agency. They may also include utopian genetic selection, with the aim of modifying human appearance and intelligence, and technological implants to expand human possibilities in many artificial ways. Other film and literary fields can be used to explore "crucial notions" like speciesism and nonhuman agency (Ferrando 2014, 169) through a posthumanist lens. These will be discussed in Chapters 8 and 9.

7.2 *The Time Machine* and the post-sapiens future

Many apocalyptic and post-apocalyptic texts represent a future without humanity. Sometimes, humans are just wiped out by a mysterious epidemic or environmental catastrophe. The few survivors risk losing the qualities that (supposedly) make them stand out as human. They may become dramatically ignorant, like the survivors of *Earth Abides* (1949, George R. Stewart) or the children of *Mad Max Beyond Thunderdome* (1985, George Miller), who do not know what a city is, or the adolescents in the next episode of the saga (*Mad Max Fury Road*, 2015, George Miller), who do not know what a tree is. They may lose their moral values or agency, as many characters in The Road and all the human survivors in Wall-E. They may even lose the position of hierarchical superiority over the species they presume to dominate and see their condition reversed (as in the novel *La Planète des singes*, 1963, by Pierre Boulle, and the many films inspired by it). Alternatively, they can evolve into post-human species.

This last scenario is at the center of *The Time Machine* (1895, by H.G. Wells). It is the story of an unnamed man, called the Time Traveler, who, thanks to a machine he has invented, travels through time and discovers a future in which different beings have replaced humans. At his first stop in the future, the Time Traveler arrives at the year 802,701. He gets to the site where London used to be, but instead of the city, he finds a decaying place inhabited by a new kind of beings, the Eloi. The Eloi look like humans, are good-looking, albeit childlike, and of limited intellect. They do not need to work. As if they were in a Golden Age, they spend their time playing, relaxing, bathing in the river. Thus, the Time Traveler initially ascribes the beauty, mild character, and minute dimensions of the Eloi to natural evolution: the ultimate victory of human beings in the struggle with "nature" must have led them to lose their strength, transforming them into childish and asexual beings. This "evolutionary" hypothesis does not last long. The Eloi live on the surface but a different and much less friendly kind of beings, the Morlocks, live beneath the ground. While the Eloi are pleasant and agreeable, the Morlocks are ugly, slimy, and white, due to their underground existence. On account of a supposedly old practice, the Morlocks provide for the Eloi's clothing and necessities, but also feed on them because they are cannibals. The Time Traveler theorizes the two peoples' roles: the Eloi are "mere fatted farm animals" for the Morlocks. Conversely, the Morlocks are, in Wells' definition, "inhuman sons of humans." They are not humans anymore, but, at the same time, they are disturbingly similar to human beings.

In the novel, post-sapiens evolution continues beyond Eloi and Morlocks: after the Traveler's first stop in time, he goes further in the future to find himself at a beach, where he is attacked by giant crabs amid "abominable desolation." Later, he arrives at another moment in time when there are no signs of life left on Earth except for a mysterious black being with enormous tentacles. In this way, Wells presents to readers different mechanisms of parallel, or different, evolution (Parrinder 2015). In his vision, human beings may not remain "one species" but degenerate at least "into two distinct animals" and turn into two different post-human species, and even disappear, being substituted by other beings.

In the late nineteenth century, Wells was not alone in this fear. Concern about a negative evolution, and thus a degeneration of the human species, was common among the intellectuals of the time, in connection with the spread of Darwinian thought and, more generally, with the idea that "the evolutionary process might be reversible" (Scherr 2019). Wells does not offer many explanations to justify the negative evolution of the human species. However, the division into two different species seems to suggest the fear that humanity will degenerate, exaggerating two potential trends: on the one hand, the loss of "manliness," and, on the other, the excess of aggressiveness. When the Time Traveler returns to Victorian London and says, "I'm starving for a bit of meat," this desire brings him closer, albeit metaphorically, to the cannibals of the future (Scherr 2019).

7.3 Transmogrifying epidemics and new world orders

In addition to evolution, other causes may transform humans into something different. For example, the hypothesis that mysterious diseases, or genetic manipulations, can spread in pandemic form, and transform humans into "walking dead" is the basis of a very successful narrative trope, the zombie apocalypse. In this case, the plot typically involves that the recently dead return to life to feed on the living, who in turn die and return to feed on others who are alive. The mechanism quickly turns almost all humans into unconscionable, voracious beings. Meanwhile, the narrative portrays a small group of people who have met by chance after the catastrophe. They develop into a social nucleus that fights the multitudes of aggressive "undead" while also trying to reorganize the disrupted social order.

This kind of narrative finds its forerunner in Richard Matheson's novel *I Am Legend* (1954). In this, due to an inexplicable disease, all human beings have been transformed into blood-sucking beings that only move at night, like vampires. Only one man, called Richard Neville, survives; he tries to carry on his life in a very "human" way, listening to music, cooking, decorating his house, but, at the same time, he is a fierce vampire fighter. After a few years, he discovers that, due to some mutations, there are two kinds of "vampires." Besides the ones that move like "robots," whom he kills without mercy, there are the "post-humans," who are infected but have overcome their disease and are trying to build a new society. Even if they do not attack him to eat him, as the former do, infected mutants also pose a threat. They hate him because they accuse him of being a vampire exterminator. In the new post-human order, Neville is therefore the "abnormal" one, the feared villain. Consequently, the post-human mutants first send Ruth, a good-looking mutant, to seduce and capture him, whereupon they sentence him to death. Before dying, he understands that post-human society members will remember him as an old "superstition," a "legend" from the past. "Robert Neville is not legend because he represents human society, nor because he somehow saves humanity; Robert Neville is a legend because he is the single largest threat to a new society, one that has superseded humanity" (Christie 2011, 68). In this way, the novel addresses an important issue: is humanity meant to survive, or is its destiny to give way to other species of post-human beings? (Koenig-Woodyard 2017).

Matheson calls Nelville's opponents "vampires," not zombies; the protagonist of the narrative is an individual, not a group, and the future of humanity is utterly without hope. Nevertheless, the novel and the first film adapted from it (*The Last Man on Earth*, 1964, Ubaldo Ragona and Sidney Salkow) are generally regarded as the inspiration for the subsequent horde of zombie texts (Christie 2011), starting with the cult movie *Night of the Living Dead* (1968, George H. Romero). Due to the ensuing proliferation of films, video games, novels, graphic novels, television series and even zombie parodies (Platts 2013), the zombie apocalypse trope is now so popular as to be "a tenacious part of mainstream American culture" (Bishop 2010, 12).

Its enormous success is linked to many factors. The greed of the zombies, whose only impulse is to eat, can be considered a metaphor for contemporary consumerism (Harper 2002), as in George H. Romero's *Dawn of the Dead* (1978). Then, there is the fear of invasion (Saunders 2012). In particular, in the post-September-11 United States, zombies have been a metaphor for Islamic terrorists, undocumented migrants, and all those who can bring contagious diseases from the outside. In that regard, an invasion of zombies lends itself to exploration of the relationship between "us" and "them," and the construction of "otherness" (Lauro 2017), not only for international relations but also concerning home policy, the political and social reorganization of human survivors, and the various modes of managing power.

Post-apocalyptic social relations usually set the main characters, their families and friends in opposition to "bad" humans, greedy and ruthless, with the "walking dead" caught between them. In the television series *The Walking Dead* (2010–, AMC), the survivors are, or risk being, just as greedy as the zombies, and even the main characters are just as aggressive (for this reason, at the end of the second season, Rick, the protagonist, says: "We are all infected") (dell'Agnese 2014a). They fight only for themselves, their families, and friends. The value attributed to interpersonal ties (family, friendship) is contrasted with that associated with external ones, in a metaphor of the neoliberalist belief that benevolence can blossom in private life, but the outside world remains a place of fierce competition (Kunkel 2008).

In addition to political considerations, zombie apocalypses may allow some environmental observations in relation to the epidemic's causes and the post-apocalyptic setting. Sometimes, the epidemic remains unexplained, as in Matheson's novel, in the first film adapted from it (*The Last Man on Earth*), and in *The Walking Dead*. Often, it has an anthropic reason (which can generally be related to the anxieties of the moment). The second movie derived from *I Am Legend* (*The Omega Man*, 1971, Boris Sagal) attributes the plague to a bacteriological war, and the third (*I Am Legend*, 2007, Francis Lawrence) to the genetic manipulations of the measles virus. In *28 Days Later* (2002, Danny Boyle) and its sequel *28 Weeks Later* (2007, Juan Carlos Fresnadillo), genetic manipulation is implicated again, while in *World War Z* (2013, Marc Foster) the cause of the epidemic is not known; however, the novel from which the movie is adapted (*World War Z: An Oral History of the Zombie War,* 2006, Max Brooks) points to possible experiments made in China. Anthropogenic involvement characterizes many of the productions between the 1970s and the early 2000s. Later, the human factor is again questioned. From this point of view, the post-apocalyptic game *The Last of Us* (2011) takes the process of human re-dimensioning further. The end of human civilization is not brought about by an unknown cause, nor is it the consequence of human action; it is the result of the biological affirmation of a more vigorous species, the parasitic fungus *Cordyceps* (which exists), which takes possession of human beings and makes them its carrier.

Generally speaking, in zombie apocalypses, the environment remains substantially unchanged or blooms again thanks to the planet's ability to reconstitute life. In each season of *The Walking Dead* television drama, the context becomes ever greener (dell'Agnese 2014b); in the videogame *The Last of Us*, the setting is marked not only by a progressive resurgence of vegetation, but also by the contrast between the quietness of the renaturalized areas and the threat posed by dilapidated human constructions (Green 2016; Fraser 2016). In this manner, the two productions convey a biocentric approach.

7.4 Improving humanity? The (anti)utopian dream of perfection

The transformation of human beings is not always bound to degenerate the species, resulting in pleasant but weak-minded creatures, monsters, or zombies. It may also be that human societies of the future improve (or try to improve) the human species, making post-human beings healthier, more beautiful, and even more long-lived. The idea is taken up critically by anti-utopian literature, which in various ways expresses concern about the progressive control of human bodies and minds.

In *Brave New World* (1932, Aldous Huxley), giving birth naturally is considered highly uncivilized. People are no longer born from a mother into a family but produced in laboratories through artificial insemination and raised in hatcheries. They are medically conditioned to have the appropriate physical and mental qualities to occupy a given social and economic position. When they are born, they are already separated into categories (Alpha, Beta, Gamma, Delta, Epsilon) so that they might be conditioned to fit the role they will have in life. The future of each individual is planned on the basis of the characteristics attributed to him or her during artificial gestation. Only Alphas and Betas are unique individuals, created by selecting biologically superior ova and fertilizing them with biologically superior sperm. Gamma, Delta, and Epsilon classes are instead produced *en masse* through a cloning process (called the Bokanovsky Process). Alphas and Beta are destined to perform skilled and intellectual jobs, while Gamma, Delta, and Epsilons will perform menial jobs for which a lower intelligence is sufficient.

"Obsession with body perfection" (Gallardo Torrano 2007) lies at the heart of the movie *Gattaca* (1997, Andrew Niccol). The film is set in a "not too distant future," when there will no longer be a place for individuals who are not genetically programmed as "perfect" beings. As in *Brave New World*, human beings are produced through the selection of physical and intellectual characteristics, but a child's physical appearance is chosen by his or her prospective parents, not by the system. Genetic manipulation permits the elimination of all negative characteristics so that everyone will be a "valid," that is to say a genetically modified and supposedly perfect individual. There are still people from older generations and a few children who are traditionally conceived, and thus not perfect, so they are called the "in-valids." According to a class division based on genetic castes (Gallardo Torrano 2007), the "valids" are selected for better professions, while the "in-valids" can perform only menial jobs. However, not even the "valids" are always perfect, or rather, not unalterably perfect. The film tells of the friendship and solidarity between Vincent, a young man who was traditionally conceived but has the ambition to be an astronaut, and Jerome, a young man who was genetically conceived, but who had an accident and is now a person who uses a wheelchair.

Genetic engineering is associated with power in another dystopian film by the same director: *In Time* (2011, Andrew Niccol). In the year 2169, together with discrimination and beauty, a third determinant is "age." To avoid over-population, people are genetically programmed to stop aging at 25, whereupon they die. To prolong their lives, they need to in some way obtain time. Time is a currency and can be earned by working but it can also be stolen or traded. As it is a currency, time is used to buy goods and even to cross the border of a district of residence. Consequently, just as in a classic capitalist system, some people accumulate time so that they can remain perpetually young and beautiful, while others die young, because "for a few to be immortal, all the others must die." When seen from this viewpoint, *Gattaca* is considered a "good example of criticism against transhumanism" (Adorno 2010, 348), while *In Time* not only provides a glimpse of "the ethical debate of human enhancement" (Roduit, Eichinger, & Glannon 2018), it also raises issues about class division and spatial segregation (Gomez Munoz 2016).

The control of bodies is associated with mind control in *Uglies*, a series of young adults novels (by Scott Westerfeld Often, 2005–18). Here too, the society of the future is stratified in accordance with bodily perfection, thus, at the age of 16, everyone undergoes an obligatory operation called The Surge to make them "beautiful." Unfortunately, together with ugliness, the operation also takes away their ability to think independently.

The theme of genetic engineering, aimed at creating post-human beings that surpass the human species, is central to Margaret Atwood's *MaddAddam* post-apocalyptic trilogy (*Oryx and Crake*, 2003; *The Year of the Flood*, 2009; *MaddAddam*, 2013). In this world, the Crakers, the "children" produced by the genetic engineer Crake, are like the Eloi in Wells' novel *The Time Machine*, who were good-looking but not very bright. The Crakers are the result of "Paradice," the name of one of Crake's projects. They are supposed to be the prototypes for the children produced to be sold to parents willing to pay for faultless progeny. They are instead a "para-human population" with which Crake plans to substitute humanity (Mosca 2013). They are beautiful and beardless with green eyes and a beautiful skin that does not burn in the sun (unlike Snowman, the human protagonist of the novel). They live on kudzu, an invasive species of vine, do not need animal proteins (they are "monstrous vegans," as Quinn 2018 calls them), and have sex only to reproduce. They are, therefore, "sustainable" yet devoid of passion: they must not have feelings, greed, religion or curiosity (even though they ask Snowman a lot of questions as from the start of the first novel in the trilogy, and, at the end of the third, they even learn how to read and write, becoming in this way more "human") (Mosca 2013).

7.5 "I'm not machine, not man. I'm more": *Terminator* and the other transhumans

The dream of improving humanity is also typical of transhumanism. In this case, however, the hope is to enhance human capacities with technology rather than with genetics. Technology also makes it possible to imitate or transform the human body. Producing a machine that looks and acts like a human means creating an android, while enhancing human bodies with technology means creating a cybernetic organism (cyb-org) (van Riper and Bowdoin 2002). In the first case, the goal is to produce artificial beings that imitate humans to a greater or lesser degree; in the second, the human body is enriched with artificial or mechanical parts to enhance its capacities. Any organism can be transformed into a cyborg, not only a human being; still, we generally refer to humans when we talk about cyborgs. Similarly, it can be assumed that any living thing can be artificially created (see the owl in Blade Runner, 1982, Ridley Scott). However, the term android refers to a robot designed to look and act like a human.

Despite this difference, literature often uses the two terms synonymously (Laughlin 1997, 144). So, for instance, Rhys Owain Thomas writes that: "Cyborgs are created either through the (often brutal) insertion of machine components into a previously wholly organic entity, as in the case of the *Star Trek* franchise's horrific Borg, or are designed from the outset as a synthesis of the organic and the artificial, such as Arnold Schwarzenegger's quintessential cyborg villain/hero in the *Terminator* movies" (Thomas 2015, 57). Nevertheless, if the previous distinction is to be maintained, Terminator is an android, and also "the remarkable movie *Blade Runner*, frequently mentioned in reference to cyborgs, really involves androids" (Laughlin 1997, 157).

Human-like machines in science fiction and dystopic literature are recurrent, both in the form of automata (like the one in *Metropolis*, 1927, Firtz Lang) and androids (as in the collection of short stories *I, Robot*, written by Isaac Asimov and published in 1950, or in *Do Androids Dream of Electric Sheep?*, 1968, by Philip K. Dick). Some literary characters, dating back to the end of the nineteenth century, are the result of a combination of a machine and a body. For instance, in *The Ablest Man in the World*, a short story by Edward Page Mitchell (1879), the character behind the title is Baron Savitch, a man with a mental disability, whose brain has been replaced by a machine, making him infallible. In contrast, *The Clockwork Man* (1923), a novel by Edwin Vincent Odle, depicts a future world in which the brains of human beings receive special clocks that allow them to travel through time and space, and thus it is considered the first sci-fi text featuring a cyborg.

However, the term cyborg was only coined in the 1960s, and with specific reference to space exploration. The inventors of the concept, Clynes and Kline (1960, 26) affirmed that, while in the past, "evolution brought about the altering of bodily functions to suit different environments … starting as of now, it will be possible to achieve this to some degree without alteration of heredity by suitable biochemical, physiological, and electronic modifications." The idea was to adapt an individual to a new context (space) rather than create an environmental bubble around him or her, enhancing the individual's body with technological aids to allow survival in the new environment. The replacement of human body parts (such as dentures and heart valves) has existed for a long time, but a cyborg is much more complicated. The cyborg idea does not envisage that the mind-body separation is maintained, i.e., that a human mind can be installed in an artificial and mechanical body and thus continue to function. It recognizes that "human consciousness and culture are functions of the human nervous system" (Laughlin 1997, 149), and that there is a reciprocal influence.

In fiction, the first literary appearance of the term is in the novel *Cyborg* (1972, by Martin Caidin), on which the American television series *The Six Million Dollar Man* (1974–78, ABC) was based. In this case, the protagonist, Steve Austin, suffers a severe accident in which he loses an eye, an arm, and both legs. The replacement of the lost limbs with bionic parts gives him extraordinary powers. Since then, cyborgs have multiplied. The most famous and celebrated are perhaps Darth Vader, from the *Star Wars* franchise (1977–), *RoboCop*, a fictional policeman created for the movie of the same name (1987, Paul Verhoeven), which then generated a multimedia franchise of a television series, graphic novels, cinematic sequels, and video games, and the most recent characters of Immortan Joe and Furiosa in *Mad Max Fury Road* (2015, George Miller).

Cyborgs can be "monsters" (Haraway 1985), that is, humans not conforming to the "bodily norms" (Shildrick 2002, 2), as the addition of any parts to the human body may alter its external appearance. Prosthetic devices can help develop basic skills essential to a cyborg's existence, or that in some way exceed human capabilities. In Western popular culture and the Hollywood film tradition, a character with a disability is generally attributed a humorous, pathetic, frightening, or antagonistic function (Safran 1998; dell'Agnese 2011). It is, therefore, not surprising that cyborgs often play the role of the villain. "In fact, cyborg characters in science fiction usually have been objects of fear (e.g., the Borg in *Star Trek: The Next Generation* [1987–94], or Damon Knight's Jim in *Masks* [1968]) and are often the bad guys in the tale" (Laughlin 1997, 155). The same is true of Darth Vader and Immortan Joe. However, not all cyborgs are bad (see *RoboCop*), and as will be seen in relation to the *Terminator* franchise, the figure of the cyborg has undergone a positive evolution.

Androids are usually "normate," in line with contemporary visual culture and Hollywood aesthetics (Garland Thompson 2017). Being created as "artificial slaves" (LaGrandeur 2015) to meet human — including sexual — needs, they are sometimes also physically attractive, creating greater ambiguity. They have been portrayed as villains, especially by the cinematography of the 1970s and 1980s, as a metaphor for the technological threat and humans' inability to control it (see for instance Ash in *Alien*, [1979, Ridley Scott], or T-800 in the first episode of the *Terminator* saga). As they can be good-looking, as well as capable of cognitive thinking and even love, androids can also become "unconventional protagonists" for whom the audience feel sympathy (Surkan 2004, 115), for example, as represented by David, the childlike android in *A.I.* (2001, Steven Spielberg). This pattern can be seen in the passage from the Gunslinger in *Westworld (*1973, Michael Crichton) to the androids in the first season of the *Westworld* television series produced about 50 years later (2016–). The two scripts share the same location and the same basic concept: androids are the attraction of an amusement park on the theme of the Far West where they play the part of frontier characters, and human tourists can treat them in any way they choose. They may have sex with them, or make friends with them, or even kill them. For each new group of tourists, the damaged or "dead" androids are repaired and returned to service. If the basic concept is the same, the perspective is different, since in the first version of *Westworld*, the androids rebel and thus play the part of the villains, while in the second version, although the androids rebel, the villains of the story are the tourists and the creators of the androids: the former because they enjoy seeing the death and blood of the androids, and the latter because they continually allow the androids to suffer, even though they know that the androids have sensitivity.

Either good or bad, cyborgs and androids are "monsters," if not for their physical appearance, for their hybrid quality. So, as pointed out by Haraway (1985), they "throw new light on the relation between the feminine and the monstrous, where both are other to the masculine subject of modernity" (Shildrick 2002, 123). Thus, popular fiction often assigns them the function of "crossing borders," that is, of questioning the boundaries of the human being, and the possibility of establishing distinctions between individuals and species, nature and culture, natural and artificial, organic and machinic.

The question of what it means to be human in *Blade Runner* does not find an answer. Here, the replicants are bioengineered beings, "more human than humans" as suggested by the advertisement, who work as slaves in off-world colonies. They have a lifespan of four years. When they mutiny and escape, they must be "retired" (to use the euphemistic expression of the movie). Nevertheless, some of them return to Los Angeles, to ask Tyrell, their "creator," to extend their lifetime. Their elimination constitutes the core of the plot; distinguishing a replicant from a human being is, in fact, increasingly tricky. Replicants should not have emotions, but they have developed consciousness. Deckard, the protagonist, is a blade runner, i.e., an android hunter and must "retire" the rebellious replicants. He breaks the rules when he falls in love with Rachael, who is herself a replicant without knowing it. Later, the movie casts doubt on whether or not Deckard himself is also a replicant. A crucial point in the film is his confrontation with Roy, the leader of the replicants. Aware that his life is coming to an end, Roy expresses a very human fear of death and the regret that his memories will fade away like "tears in rain."

The relationship between organic and machinic is also central to the *Terminator* cinematic saga: *The Terminator* (1984, James Cameron); *Terminator 2. Judgment Day* (1991, James Cameron); *Terminator 3. Rise of the Machines*

(2003, Jonathan Mostow); *Terminator Salvation* (2009, McG); *Terminator Genisys* (2015, Alan Taylor); *Terminator: Dark Fate* (2019, Tim Miller). Here, the core of the story is the clash between humans and the machines, dominated by Skynet, a network of computers that have become so intelligent that they have taken control of the world. Through the episodes of the saga, a transition occurs from the classic human–machine confrontation, in which the machine is the villain, to an investigative perspective on the complicated relationship between the two sides. The first film is set in Los Angeles in 2029 and sees the machines send a T-800 terminator (a war robot with human features) back in time to eliminate Sarah Connor, the future mother of John Connor, who in turn is the future leader of the Human Resistance. T-800, played by Arnold Schwarzenegger, looks human but is indeed an android powered by electronic circuits. The plot, therefore, still revolves around the absolute otherness of machines, and this otherness allows the humans to assume "any ethical responsibility for future catastrophe" in "a modernist critical perspective" (Brereton 2005, 199–203). In the second episode of the saga, having failed to kill Connor's mother, the machines send a more sophisticated robot (T-1000) back in time to kill John Connor as a teenager, while the humans send a second T-800 Terminator, which looks the same as the first, but has been modified in order to protect Connor. Here, the situation is more nuanced, since T-800 is a "technically outdated," though good robot (a role it maintains also in *Terminator 3*), while T-1000 is a machine totally devoid of feelings. The boundary between good and bad, human and artificial, is blurred in the fourth film (*Terminator Salvation*), where the character of Marcus Wright is included alongside John Connor. Marcus was a man, but after being sentenced to death, was turned into a "Terminator Human Hybrid," a cyborg with a human brain, but a much stronger and faster body. "It's real flesh and blood… the heart is human and very powerful, the brain too, but with a chip interface; it has a hybrid nervous system. One human cortex, one machine," as explained to John Connor. Marcus, who is still convinced he is human, breaks the rules imposed by Skynet when he takes the side of the resistance and sacrifices himself to save Connor's life. Like Rachael in *Blade Runner*, he is somehow anti-Cartesian (Roden 2015), which means that he thinks, but that this alone does not make him human ("What it is that makes us human?" a voice-over explains at the end of the movie "It's not something you can program. You can't put it into a chip. It's the strength of the human heart. The difference between us and machines."). In the fifth episode, a variant of John Connor is infected by Skynet with nanotechnology, becoming a cyborg; in the trailer of the movie, he declares "I'm not machine, not man. I'm more," making a clear reference to the transhuman perspective. In the fifth sequel, which does not follow the plot of the previous episodes, Sarah Connor is joined as the main character by Grace, a cybernetically enhanced woman-soldier. The transition of the transhuman character (android or cyborg), from the villain of the origins to the main character and hero, is here completed (Hauskeller, Philbeck, & Carbonell 2015, 4).

Bibliography

Adorno, Francesco Paolo (2010). Against Posthuman Ideology: Aesthetics and Finitude of the Individual. *RES: Anthropology and Aesthetics* 57(1), 344–354. https://doi.org/10.1086/RESvn1ms25769988

Agnew, John A. (2003). Geopolitics: Re-visioning world politics. Routledge.

Bishop, Kyle William (2010). *American Zombie Gothic. The Rise and Fall (and Rise) of the Walking Dead in Popular Culture*. McFarland& Co.

Braidotti, Rosi (2013). *The Posthuman*. Polity Press.

Brereton, Pat (2005). *Hollywood Utopia: Ecology in Contemporary American Cinema*. Intellect Books.

Christie, Deborah (2011). A Dead New World: Richard Matheson and the Modern Zombie. In Christie, Deborah, & Lauro, Sarah Juliet (eds.), *Better Off Dead: the Evolution of the Zombie as Post-Human* (pp. 67–80). Fordham University Press.

Clynes, Manfred E., & Kline, Nathan S. (1960). Cyborgs and Space. *Astronautics*, 26–27.

dell'Agnese, Elena (2011). Mens sana in corpore sano? dis-abilità e «differenza fisica» (*) fra Hollywood e Bollywood. *Bollettino della Società geografica italiana* 4(1), 113–124.

dell'Agnese, Elena (2013). The Political Challenge of Relational Territory. In Featherstone, David, & Painter, Joe (eds.), *Spatial Politics: Essays for Doreen Massey* (pp. 115–124). Wiley.

dell'Agnese, Elena (2014a). The Walking Dead. Siamo tutti infetti? In Amato, Fabio, & dell'Agnese, Elena (eds.), *Schermi americani. Geografia e geopolitica degli Stati Uniti nelle serie televisive* (pp. 39–52). Unicopli.

dell'Agnese, Elena (2014b). Post-Apocalypse Now: Landscape and Environmental Values in The Road and The Walking Dead. *Geographia Polonica* 87(3), 1–16.

Dendle, Peter (2008). Monsters, Creatures, and Pets at Hogwarts: Animal Stewardship in the World of Harry Potter. In Heilman, Elizabeth E. (ed.), *Critical Perspectives on Harry Potter* (pp. 175–188). Routledge.

Ferrando, Francesca (2013). Posthumanism, Transhumanism, Antihumanism, Metahumanism, and New Materialisms. *Existenz* 8(2), 26–32.

Ferrando, Francesca (2014). Posthumanism. Tidsskrift for kjønnsforskning 38(2), 168−172.

Ferrando, Francesca (2015). The Body. In Ranisch, Robert, & Sorgner, Stefan Lorenz (eds.), *Post- and Transhumanism: An Introduction* (pp. 213–226). Peter Lang.

Fraser, Emma (2016). Awakening in Ruins: The Virtual Spectacle of the End of the City in Videogames. *Journal of Gaming & Virtual Worlds* 8(2), 177–196. https://doi.org/10.1386/jgvw.8.2.177_1

Gallardo Torrano, Pere (2007). The Body as Utopia: Gattaca, by Andrew Niccol (1997). *Spaces of Utopia* 5, 44–54.

Garland Thompson, Rosemary (2017). Introduction: From Wonder to Error. A Genealogy of Freak Discourse in Modernity. In Garland Thompson, Rosemary (ed.), *Freakery: Cultural Spectacles of the Extraordinary Body* (pp. 1–19). New York University Press.

Gomez Munoz, Pablo (2016). Keeping Workers at a Distance: The Connection Between Borders and Finance in Andrew Niccol's In Time. *Geopolitics* 21(1), 195–214. https://doi.org/10.1080/14650045.2015.1132704

Gray, Chris Hables (2017). Post-Sapiens: Notes on the Politics of Future Human Terminology. *Journal of Posthuman Studies* 1(2), 136–150. www.jstor.org/stable/10.5325/jpoststud.1.2.0136

Green, Amy M. (2016). The Reconstruction of Morality and the Evolution of Naturalism in the Last Of Us. *Games and Culture* 11(7–8), 745–763. https://doi.org/10.1177/1555412015579489

Haraway, Donna (1985). Manifesto for Cyborgs: Science, Technology, and Socialist Feminism in the 1980s. *Socialist Review* 80, 65–108 (also as A Cyborg Manifesto: Science, Technology, and Socialist-Feminism in the Late Twentieth Century. In Haraway, Donna (ed.), *Simians, Cyborgs, and Women*, Routledge, 1991, and in Haraway, Donna, *Manifestly Haraway*, University of Minnesota Press, Minneapolis, 2016).

Harper, Stephen (2002). Zombies, Malls, and the Consumerism Debate: George Romero's Dawn of the Dead. *Americana: The Journal of American Popular Culture* 1(2). www.americanpopularculture.com/journal/articles/fall_2002/harper.htm

Hauskeller, Michael, Philbeck, Thomas D., & Carbonell, Curtis D. (2015). Posthumanism in Film and Television. In Hauskeller, Michael, Carbonell, Curtis D., & Philbeck, Thomas D. (eds.), *The Palgrave Handbook of Posthumanism in Film and Television* (pp. 1–7). Palgrave Macmillan.

Koenig-Woodyard, Chris (2017). "Lovie – is the vampire so bad?": Posthuman Rhetoric in Richard Matheson's I am Legend. In Heise-von der Lippe, Anya (ed.), *Posthuman Gothic* (pp. 77–92). University of Wales Press.

Kunkel, Benjamin (2008). Dystopia and the End of Politics. *Dissent* 55(4), 89–98. https://doi.org/10.1353/dss.2008.0072

LaGrandeur, Kevin (2015). Androids and the Posthuman in Television and Film. In Hauskeller, Michael, Carbonell, Curtis D., & Philbeck, Thomas D. (eds.), *The Palgrave Handbook of Posthumanism in Film and Television* (pp. 111–119). Palgrave Macmillan.

Laughlin, Charles D. (1997). The Evolution of Cyborg Consciousness. *Anthropology of Consciousness* 8(4), 144–159. https://doi.org/10.1525/ac.1997.8.4.144

Lauro, Sarah Juliet (ed.) (2017). *Zombie Theory: A Reader*. University of Minnesota Press.

Lemmens, Pieter (2015). Book review: Robert Ranisch and Stefan Lorenz Sorgner: Post- and Transhumanism: An Introduction. *Human Studies: Journal for Philosophy and the Social Sciences* 38(3), 431–438. [https://doi.org/10.1007/s10746-015-9354-4]

Mosca, Valeria (2013). Crossing Human Boundaries: Apocalypse and Posthumanism in Margaret Atwood's Oryx and Crake and The Year of the Flood. *Altre Modernità* 9, 38–52. https://doi.org/10.13130/2035–7680/2985.

Parrinder, Patrick (2015). *Utopian Literature and Science*. Palgrave Macmillan.

Platts, Todd K. (2013). Locating Zombies in the Sociology of Popular Culture. *Sociology Compass* 7(7), 547–560. https://doi.org/10.1111/soc4.12053

Quinn, Emelia (2018). Monstrous Vegan Narratives: Margaret Atwood's Hideous Progeny. In Quinn, Emelia, & Westwood, Benjamin (eds.), *Thinking Veganism in Literature and Culture* (pp. 149–173). Palgrave Macmillan.

Ranisch, Robert, & Lorenz Sorgner, Stefan (2015). Introducing Post- and Transhumanism. In Ranisch, Robert,& Lorenz Sorgner, Stefan (eds.), *Post- and Transhumanism: An Introduction* (pp. 7–27). Peter Lang.

Roden, David (2015). Post-Singularity Entities in Film and TV. In Hauskeller, Michael, Carbonell, Curtis D., & Philbeck, Thomas D. (eds.), *The Palgrave Handbook of Posthumanism in Film and Television* (pp. 88–98). Palgrave Macmillan.

Roduit, Johann A.R., Eichinger, Tobias, & Glannon, Walter (2018). Science Fiction and Human Enhancement: Radical Life-Extension in the Movie "In Time" (2011). *Medicine, Health Care and Philosophy* 21(3), 287–293. [https://doi.org/10.1007/s11019-018-9831-4]

Safran, Stephen P. (1998). The First Century of Disability Portrayal in Film: An Analysis of the Literature. *The Journal of Special Education* 31(4), 467–479. https://doi.org/10.1177/002246699803100404

Saunders, Robert A. (2012). Undead Spaces: Fear, Globalisation, and the Popular Geopolitics of Zombism. *Geopolitics* 17(1), 80–104.

Scherr, Alexander (2019). The Morlock-Eloi Illusion: Shifting Monstrosities in H.G. Wells' The Time Machine in the Context of the Degeneration Discourse. *Anglistik* 30(3), 121–133. https://doi.org/10.33675/ANGL/2019/3/11

Shildrick, Margrit (2002). *Embodying the Monster: Encounters with the Vulnerable Self.* Sage.

Surkan, Kim (2004). "I Want to Be a Real Boy": AI Robots, Cyborgs, and Mutants as Passing Figures in Science Fiction Film. *Femppec* 5(1), 114–136.

Thomas, R.O. (2015). Terminated: The Life and Death of the Cyborg in Film and Television. In Hauskeller, Michael, Carbonell, Curtis D., & Philbeck, Thomas D. (eds.), *The Palgrave Handbook of Posthumanism in Film and Television* (pp. 57–65). Palgrave Macmillan.

Van Riper, & Bowdoin, A. (2002). *Science in Popular Culture: a Reference Guide.* Greenwood Publishing Group.

Wolfe, Cary (2010). *What is Posthumanism?*. University of Minnesota Press.

8 Viewing dogs with (post)human lenses

8.1 "Dogs are my story here, but they are only one player in the large world of companion species" (Haraway 2003, 25)

Cyborgs and dogs, according to Donna Haraway (2003), have something in common. "Each brings together the human and non-human, the organic and technological, carbon and silicon, freedom and structure, history and myth, the rich and the poor, the state and the subject, diversity and depletion, modernity and postmodernity, and nature and culture in unexpected ways" (Haraway 2003, 4; 2016, 96).

Sci-fi and dystopian narratives provide an excellent stage to investigate posthuman and transhuman perspectives in popular culture, and to see how they try to answer the question about what it means to be human. From a different point of view, the posthuman approach rejects "the humanist belief that 'man is the measure of all things'" (Ranisch and Sorgner 2015, 16). So, it raises other crucial issues about the position of humans in their relationship with nature within an array of narratives, to understand whether the nature–culture dualism is undone, or whether, on the contrary, humans are still "exceeding other creatures in their control over nature" (Fuller 2015, 204). Therefore, the analysis is extended to other literary fields while keeping one subject, the dog. Dogs and humans are bonded in "significant otherness" (Haraway 2003, 2016), and have a "cohistory" with humans as "companion species." For this reason, to quote Haraway again (2003, 25), they are "my story here, but they are only one player in the large world of companion species." Studying how dogs are represented in popular culture is only one way to see if animal representation is "something which encourages engagement with the experiences of animals," or instead makes them merely "representational resources that can be employed for the purposes of rendering legitimate human–animal hierarchies" (Mills 2017, 71).

In real life praxis, speciesism and anthropocentrism are still the dominating discourse. Most of the time, dogs are victims, subjugated to the power of their master. The brutality of the power exercised by the human being over the dog is already highlighted by the German philosopher Schopenhauer, in the mid-nineteenth century: "man's most loyal friend, the very intelligent dog, he puts upon a chain! I can never look at such a dog without heartfelt compassion for it and deep indignation for its master, and with gratification I think of the case reported a few years ago by The Times, of a certain Lord who kept a large dog on a chain. Once, strolling through his yard, he could not resist the urge to pet the dog, whereupon it immediately tore open his arm from top to bottom – justifiably! It was trying to say: 'You are not my master, but my devil, who makes my brief existence into a hell.' May this happen to all who keep dogs on a chain" (*Parerga and Paralipomena: short philosophicalessays*, 1851).

The dog's place is at the service of the human being, and success depends on the ability to play by the rules and the benevolence of the human master. As the feminist Charlotte Perkins Gilman wrote in 1911, "no creature has been more constantly under man's [sic] hand than the dog." Then she goes on: "A slave without any industry to justify his slavery; a prisoner, for no fault to warrant his imprisonment; a captive, led in chains and manacled in his one point of contact with life, his means of inquiry, of expression, of defense, of eating, breathing and panting—his poor muzzle—this is the animal we say we love!" (Perkins Gilman 1911, 180–182). Indeed, the relationship human–dog, far from being equal, may mix in different doses "dominance and affection" (Tuan 1984), but it always assigns power to one side only, inside an inherently unequal and "patriarchal" relations.

Pedigree dogs, though usually privileged, are in no better position. As remarked by Tuan (1984), breeding animals for aesthetic purposes is an example of how humans enjoy manipulating animals' life. Charlotte Perkins Gilman, in a similar vein, but with more crude language, speaks of "the little deformities so carefully developed to please our refined tastes," and adds: "There will come a time when a cultivated civilized human being will feel only pain and horror at the distortion of a live animal, either by the simple savagery of mutilation or the more refined, insinuating cruelty of artificial breeding" (1911, 206–209). H.G. Wells (1928) expresses a similar attitude when speaking about "fancy" dogs as "invalid and grotesque deformations of the canine type," and of dog breeds as an "ugly trade in animal deformity."

These accusations are, though, isolated voices within a vision that instead tends to present a submission to the master (symbolized by leash and collar) as the only positive outcome of canine life. The dog's position, vacillating between an object of power and an object of love, is typically present in popular culture, although, in general, dominance prevails over affection. As a result of their "cohistory" with humans, dogs are ever-present in Western popular culture (McHugh 2004), starting perhaps with *The Odyssey*, where the old hunting dog Argos is the only one who recognizes the Homeric hero on his return home and promptly dies after it. Dogs are the subjects of poems, novels, movies, television series, and even video games (*Dog's Life*, 2003, Frontier Developments; *Nintendogs*, 2005, Nintendo). Sometimes dogs are just mentioned as extras, and sometimes they play the role of the main character. They are supporting actors in detective stories, standing beside the (male) main to symbolize loyalty (like

Alfie, the Old English Sheepdog in *Serpico*, 1973, Sidney Lumet; Tequila, in the *Tequila and Bonetti* television series, 1992, CBS; Jerry Lee in the movie *K-9*, 1989, Rod Daniel and sequels; Hooch in *Turner and Hooch*, 1989, Roger Spottiswoode; Rex, in the Austrian television series *Kommissar Rex*, 1994–2004). Alternatively, they play a symbolic role (again standing alongside the male protagonist). In dystopian/post-apocalyptic narratives, to have a dog means having emotional ties, as in *Earth Abides* (1949, George R. Stewart), *Mad Max 2: The Road Warrior* (1981, George Miller), *I Am Legend* (1954, Richard Matheson), and the television series *The Walking Dead* (2010-on, AMC). They may appear on the screen for only a few seconds but still work as a cultural signal; in *The Road* (2006, Cormac McCarthy), having a dog means to have family values, and so only the family met by the Son at the end of the narrative, which has a dog, qualifies in this way. There are no dogs in *Blade Runner* (1982, Ridley Scott), but the idea of the dog still resonates symbolically as an ethnic boundary (Barth 1969). When Deckard, the replicant hunter, submits Rachael to the Voigt-Kampf test to understand if she is a replicant, he understands that she is an android because she fails to react with the expected emotion to the image of an entrée consisting of a boiled dog, thereby revealing her essence (Molloy 2011).

Whether they are at the center of the scene or just extras, dogs may be portrayed in many different ways. They can be heroes, like Rin Tin Tin, the famous German shepherd brought to the US from Germany after the First World War, which featured in 29 films and was turned, after his death, into a brand for dozens of films and television series with his name; or Balto, the Siberian Husky celebrated by an animated film in 1995, that dragged a sled load of dysentery medicines in a race against time in Alaska in 1925. Dogs may also be "good girls," always coming home, like *Lassie*, the Rough Collie from Eric Knight's novel of 1938, which was later celebrated in about ten movies, a television series (1954–73), and even a radio show (1947–50). They can be adventurous seducers (like the Tramp, in *Lady and the Tramp*, 1955, Walt Disney), disastrous troublemakers (like *Beethoven*, the Saint Bernard in the film series of the same name, 1992–2014), or even terrible dogs like *Cujo* from the horror novel by Steven King, 1981, though he only turns murderous because he is infected with rabies.

Sometimes a dog breed is promoted as a fashion (Rough Collies after *Lassie*; Dalmatian dogs after *One Hundred and One Dalmatians*, 1961, Walt Disney; Old English Sheepdogs after *Serpico*, Chihuahuas after *Beverly Hills Chihuahua*, 2008, Walt Disney). The legitimate existence of breeds, conversely, is hardly ever called into question by popular culture. Some ironic hints can be found in Virginia Woolf's novel, *Flush: a Biography* (1933); when the cocker spaniel Flush arrives in Italy, he notes that it is hard to see purebred dogs around, only crossbreeds. Feeling himself "the sole aristocrat among a crowd of canaille," he wonders: "Had the Kennel Club,[1] then, no jurisdiction in Italy? Was the Spaniel Club unknown? Was there no law which decreed death to the topknot, which cherished the curled ear, protected the feathered foot, and insisted absolutely that the brow must be domed but not pointed? Apparently not" (*Flush*, 1933).

A critical attitude towards canine breeds is also evident in another story by Woolf, *Gypsy, the Mongrel* (1940), where the comparison between the two dogs of the story, Gypsy, the little crossbreed full of personality, and the dull setter Hector, makes the author's attitude clear (Reynier 2009). Gypsy, the mongrel, has no patience with rich and spoiled dogs, such as pugs and other fancy breeds. As a symbol of the spoiled dog, the pug returns in the animated movie *Pocahontas* (1995, Walt Disney). In this case, the comparison between Percy, the pug belonging to Governor Ratcliffe, and the little wild animals surrounding the Native American girl, highlights the "artificiality" of the purebred dog versus the naturalness of the other animals.

Other dog breeds too are often associated with clichés, such as the centuries-old one that links Pekinese, Maltese, and small dogs in general with affluent people, especially women (McHugh 2004). For example, "lap" or "toy" dogs have become an elitist sign of nobility or a living ornament that adorns the aristocratic lady. Such symbolism may be used to make social criticism, again using dogs as a metaphor. The satirical poem *Il giorno*, written by Giuseppe Parini in 1763, describes a dialogue at an aristocratic lunch where a young and elegant lady of high society cries as she listens to the invective of a vegetarian against the cruelty of those who kill oxen and lambs for their meat. She has just fired her servant for kicking her little dog (the Virgin Cuccia) after being bitten (Parini's implication is that she cries for the lambs but does not mind having thrown the servant's family into misery). Other breeds, such as hunting dogs and greyhounds, are associated with aristocrats (males), while pit bulls are often identified "with urban, poor and specifically black people" (McHugh 2004). The mutt, on the other hand, represents the frontier character, especially in North American literature, that must be brought back under the control of a master and put on a leash (i.e., learn to respect the laws).

These many fictional dogs, on paper and screen, usually do not honor real dogs. They are anthropomorphized, gendered, stereotyped, and portrayed in an unbalanced power relationship. At times, they are used for purposes of symbolism. Very rarely are they represented as the "significant other" of humans. On a few occasions only have some authors tried to take their point of view and penetrate their *Umwelt*, their way of perceiving space and their animal uniqueness. After considering in broad outline how the vast majority of texts represent dogs and their relationship with human beings, we will examine the texts themselves.

8.2 Dogs on the leash: "the law of the stronger over the weaker"

Far from being a form of companionship, the dog–human rapport risks being naturalized as a servant–master relationship by its representation in popular culture. In addition, in this relationship of control and domination, the dog, who is always subject to the conditions imposed by human beings, risks becoming the metaphor of "nature" as if to demonstrate the full control exercised over it by culture, and confirming how anthropocentrism and speciesism are pervasive when it comes to talking about animals.

As underlined by Karla Armbruster (2002), there is a classic narrative formula in many American stories that have a dog as a protagonist: though at the beginning he is a bit reluctant to learn good manners, the dog accepts losing his "wild" side and becomes a "good dog," that is to say a faithful dog who respects the rules and is perfectly adapted to life in human society. "In this classic formula, a dog joins a human family or community and must be socialized to behave appropriately. In other words, some of its natural impulses and instincts must give way to the demands of human culture" (Armbruster 2002, 353). To demonstrate her thesis, Armbruster quotes *Old Yeller*, a novel by Fred Gipson (1956) that was adapted into a film by Walt Disney in 1957. *Old Yeller* is the story of a big dog who wins the affection of a family of pioneers, defending them from bears and raccoons, and even from a wolf, but the wolf has rabies, and poor Old Yeller must be put down because he is infected.

To this narrative structure, it must usually be added that the dog is male and a mutt. The fact that the dog is not a purebred animal is fundamental because, as a mutt, he acquires a figurative image that represents the "hybrid identities" typical of "Anglo-American masculinity during the closing of the western frontier" (McHugh 2004). John Muir's celebrated short story *Stickeen. An Adventure with a Dog and a Glacier* (1897, 1909) falls in the same narrative formula, even though it appears to overturn it. In this story, the dog does not obey orders (so much so that "none of us was able to make out what Stickeen was really good for") and is described as "odd, concealed, independent." There is no request for sacrifice, abnegation, or loyalty from the dog, not an anthropomorphic reading of the character, just the attempt by Muir to understand how the dog reacts, or at least to acknowledge in him a capacity for choice and independent action. Stickeen is a dog who never complains and who, despite being small enough to resemble a toy dog, enthusiastically puts up with any weather condition and tackles any outdoor adventure. Muir recognizes him as "a true child of the wilderness" and so, he "calls into question one of the most deeply seated ideas in Western culture, namely that there is an unbridgeable chasm of difference between human and animal consciousness" (Morris 1996, 140). The tale has been described as "a challenge to the entire humanist perspective on consciousness" (Morris 1996, 140). Karla Armbruster (2002, 364) defines the tale as an inversion of the classic formula of the "good dog," because for Muir a "good dog" is "one at home in wild nature." Still, even if in an inversion of the classical narrative structure of the "good dog," the story somehow recreates the same nature–culture tension. The dog, a male mutt, belonging to another member of the expedition to Alaska described in the tale, is initially distracted, not very affectionate, and reluctant to obey commands. After having shared a moment of great emotional tension (the dangerous crossing of a crevasse) with Muir, he becomes extraordinarily fond of him ("instead of holding aloof, he always lay by my side, tried to keep me constantly in sight, and would hardly accept a morsel of food, however tempting, from any hand but mine"). At the end of the story, Stickeen does not learn to make himself useful like other "good dogs," but he has become "a changed dog," in the sense of being much more attached to a human.

In *Beautiful Joe: An Autobiography* (1894, Margaret Marshall Saunders), the main character is Joe, a "cur," as he repeats several times at the beginning of his story. Joe is a male dog who is already quite ugly, but made even uglier by the mutilations inflicted on his ears by his first master, and for that reason is given the nickname Beautiful Joe. He is rescued from this cruelty by a family of animal-lovers to whom he shows courage and faithfulness, and becomes a very "good dog," so good as to be described as such about 20 times in the novel. He craves long races, "the delicious air, and the lovely smell of flowers and grass," yet he is happy to remain at heel, like the family's other dogs. The most welcome gift is a collar, with his name on it: "Beautiful Joe."

Both *The Call of the Wild* (1903) and *White Fang* (1906), Jack London's two famous novels, both of which were later adapted into a multitude of film versions, revolve around the "nature vs. civilization" conflict, and again, the main character in both novels is a male mutt. Both books "flirt with the dog–wolf boundary" (Onion 2009). In the first case, the dog chooses to live with wolves; in the second, the dog-wolf chooses to live with humans. The plot of *The Call of the Wild* is well known: a large dog named Buck, a cross between a Saint Bernard and a Scotch Shepherd, is stolen from his master's garden in California and sold to be used in the Klondike as a sled dog. He is passed from hand to hand, learns "the law of the club" and to assert himself over rival dogs to the point of killing one of them, finally finds a master who loves him, and when this one dies, he regains his savagery and becomes the leader of a pack of wolves on the loose. So Buck, the good family dog, undergoes a coming-of-age journey in the frozen wilds of the Klondike and eventually returns to the wilderness, though only after his beloved master has left him forever. White Fang is a wolfdog born wild in the Yukon Territory, but he is tamed and, after a thousand vicissitudes, ends up as a "good dog" (with puppies) on a sunny porch in California.

Where the North Begins (1923, Chester M. Franklin and Millard Webb) was the first starring role for the German Shepherd dog Rin Tin Tin, the canine star of the Twenties. The film tells the story of a puppy lost in frozen Alaska, raised by wolves, and then domesticated by a fur hunter. While the dog understands during their first encounter that "man is not his enemy, but his master and friend," later on he fully performs the part of the "good dog" by repeatedly saving his master's life (as written in one of the silent film's captions). Again, the happy ending requires that the wild dog, who has become a good dog, finds a partner and has a litter in his new home.

The process of denaturalizing the dog and imposing rules, to turn him into a "good dog" deserving of love, does not stop at stories set in the Old West. Popular culture also endorses the "bourgeoisification" (Howell 2000, 48) of city dogs, which must have a leash and, above all, a master. In this "domestic ideology," dogs must have a "space," a kennel, in the countryside, or a house in the city, while "canine strays, like indigent humans … [are] identified as symptoms of social problems" (McHugh 2004). The restriction of dogs "to particular spaces and places, under specific restraints and categories of existence," is significant as "symptomatic of the modern, Western culture of nature" (Howell 2012, 222). Stray dogs are not admitted; they end up at the pound and, if no one claims them, they must be eliminated. On this last point, popular culture disagrees. In the animated film *Lady and the Tramp*, the sad scene of the poor dog Nutsy, taking the "long walk" to be put to sleep in the gloomy context of the pound, communicates a clear anti-euthanasia message. Similarly, in *A Dog's Purpose* (2017, Lasse Hallström), the initial scene of the crossbred puppy, who, while playing with his siblings and wondering about the meaning of a dog's life, is captured by the animal control officers to be euthanized, cannot but draw disagreement towards the procedure.

Lady and the Tramp provides several more reasons for analysis. First of all, Tramp, the male protagonist, is a stray mutt, while the female is domestic and a homebody. Then, the movie proposes, with some variation, the classic narrative structure of the "good dog" (even if it endorses the leash but not the muzzle). Tramp is a stray with "a talent for survival" (Goldmark and McKnight 2008, 110). At the beginning of the story, he lives by exploiting the benevolence of different humans, who all call him by different names and feed him on different days of the week: thus, he is portrayed as a dog who does not comply with the rules. One day, Tramp meets Lady, a purebred spaniel who has run away from her family after someone tried to make her wear a muzzle. The pair fall in love, then return to Lady's house, where the Tramp too is well received. In the happy ending, he is finally "placed," as is appropriate, in the family of his beloved Lady and, as a sign of belonging, is given a collar. At the end of the film, they have a mixed-breed litter, three of whom look like her, one like him. The 2019 live-action adaptation of the movie (*Lady and the Tramp*, Charlie Bean) eliminated the scenes considered unsuitable for the audience's new sensitivity. Alterations were made to the scene when Lady has to be defended by Tramp against a bunch of aggressive dogs: in order to eliminate the possibility of macho gender constructions, Lady defends herself. Her mistress does not hit her anymore when she behaves improperly, she simply says "Bad Dog." Nor do Tramp and Lady have puppies as the idea of uncontrolled dog breeding is no longer appropriate. However, the gist of the story remains the same: Tramp must be integrated into a family to be a "good dog," and must be leashed in the same way that Lady is. In this way, "the classic formula tells us not just a great deal about dominant cultural expectations for dogs, but also about how we envision our own relationship to the natural world. Nature is seen as powerful, but fundamentally as an opponent, as something to be conquered and bent to our own uses" (Armbruster 2002, 360).

The English writer Ouida (the pseudonym of Maria Louise Ramé) bucks the trend with her "canine autobiography" of about 600 pages, which also has a long title: *Puck: his Vicissitudes, Adventures, Observations, Conclusions, Friendships, and Philosophies* (1870). The novel tells the story of a "thoro-bred" Maltese dog (so the dog defines himself), who faces many adventures and misadventures, is sold and stolen, and in the meanwhile finds a way to denounce all the harassment of animals by human beings. Then, he speaks about the unfair destiny of the ox, who is mutilated and then taken to slaughter, of sheep, who are tortured and bruised, of the racehorse, "caressed by princes" at the hour of victory, and then "cast off to street hire" when old, and underlines that all of them are subject to "the law of the stronger over the weaker." What matters most is that the pedigree dog overturns the issue: "One day I got out 'on the loose,' as your slang phrases it; a reprehensible practice, no doubt, but one dear to dogs as to men, for better is a bare bone in the gutter, with the sweetness of free-will, than are fatted meats eaten within the curb and the gall of a chain" (*Puck,* 1870).

8.3 *Tray* and *Trixy*: vivisection and the antivivisectionist debate

In the late nineteenth and early twentieth century, a debate on vivisection flared up on both sides of the Atlantic. It can be said, perhaps in an exaggerated way, as Wells (1928) pointed out, that "it is about the dog that the controversy centers," which led to the publication of many literary works on dogs.

Vivisection is, of course, a practice based on strongly speciesist positions; indeed, Ryder developed the very idea of speciesism with relation to the use of animals in the laboratory (Ryder 2006). The voices opposing vivisection had more articulate positions, that were sometimes anthropocentric (doctors performing vivisection can become insensitive; dogs owned as pets cannot be used for vivisection; like women, dogs too are victims of men), and at others motivated by more deeply animalistic reasons (dogs have a consciousness; there is an atrocious disparity of power between those who practice the experiment and those who undergo it).

The debate ignited the minds of intellectuals and writers, many of whom used literature to send a message, with a clear awareness of the power of representation. To this end, alongside poetry and novels written in conventional literary forms, original literary forms (such as dog autobiography) were also experimented with.

Tray and Trixy are two of these literary dogs. They are separated by an ocean (Tray is English, Trixy American), but they both risk a similar fate, the vivisectionist's table. Tray is a mongrel, the protagonist of a poem (*Tray*, 1879, in the collection *Dramatic Idylls*) by the English poet Robert Browning. He is a "good dog" who saves a child in danger of drowning and then even draws her doll from the water. For this action, he attracts the attention of a vivisector who wants to see "how brain secretes dog's soul." Trixy is a poodle and the protagonist of a novel (*Trixy*, 1904) by the American author and feminist Elizabeth Stuart Phelps. Like Buck in *The Call of The Wild*, Trixy is a stolen dog. She is not stolen for a ransom, as many dogs were in Victorian London (Howell 2000 and 2015), but to be the subject of an experiment by vivisectors. Like many other literary dogs of the time, Tray and Trixy became the banner of the fight against vivisection.

In 1824, a first Society for the Prevention of Cruelty to Animals (SPCA) had already been founded in London. Even if at that time vivisection in England was quite rare (Bates 2017), in its Founding Statement the SPCA condemned "the practice of dissecting animals alive, or lacerating, mutilating, and inflicting torture upon them in various modes, to satisfy an unprofitable curiosity" (Kean 1998, 36). As the century continued, animals became the focus of a growing awareness. A positive attitude towards animals and their welfare was accentuated by Darwin's theories, which, demonstrating the continuity between human beings and nonhuman animals, offered a rationale for greater empathy with animal suffering. However, Darwin's theories also provided an argument that invigorated the practice of vivisection: "the real growth of vivisection in Britain dated from Darwin's arguing for an understanding of the commonality between species" (Kean 1998, 97). "The effect of Darwin's ideas was both to make the human more animal and the animal more human, destabilizing boundaries in both directions" (Morse and Danahay 2017, 1).

During the second half of the nineteenth century, in the field of animal welfare a growing focus was placed on dogs, albeit not always in their favor. At the time, city streets were full of dogs; some were pets, often clearly purebred, while the rest were strays. Those that were pets lived in their owners' home. When they were outside, they wore a collar and were taken for walks on a leash. Strays were progressively excluded from the public space (Howell 2015). Purebred dogs were portrayed in paintings and carved in statues (Kean 2003), exhibited in dog shows (the first was held in 1859), and more rigorously tailored to breed standards. In 1873 the Kennel Club was created to set the rules for dog shows, determine breeds, and list pedigrees.

As the popularity of purebred dogs as pets increased, so did the theft of dogs for ransom, which rapidly became "professional" and properly organized (Howell 2000 and 2012). For the strays, the first shelter in London (Battersea Dogs' Home) was established in 1860. Initially, it was a place where stray or lost dogs could be looked after and cared for while waiting to be returned to their owners. Then, a Dogs Act was issued (July 1871) which stated that "Stray dogs may be detained and sold or destroyed." Battersea Dogs' Home was endowed with a death chamber.

Vivisection became increasingly practiced, especially on dogs picked up from the streets, or stolen and sold. The increase in the number of animal experiments, linked to the professionalization of medicine and the spread of physiology laboratories using live animals, triggered public attention, and encouraged the formation of the first abolitionist movements. Some doctors were reluctant to practice vivisection and objected to it from a moral viewpoint (Bates 2017). Beyond the mere rejection of an exercise of cruelty towards sentient beings, the antivivisectionist principles were essentially based on three elements (Farmer 1996). First, vivisection had the effect of making the practitioner morally insensitive, since "Were the feelings of experimental physiologists not blunted, they could not long continue the practice of vivisection" (Hoggan, 1875). Then, the motivations behind the practice of vivisection were often linked to "the lust for scientific knowledge," as Lewis Carroll noted in 1875, rather than to obtain useful scientific results. Finally, human beings had no right to assume superiority over the "inferior animals" that could fall prey to vivisection. Within this strand of thought, there were those like Vernon Lee (1882), who, overcoming the sentimental attitude shared by the large majority towards the "faithful dog" subjected to unjust cruelty, emphasized the extreme difference in power between the animal victim, totally deprived of agency, and the human executioner (Pireddu 2014).

Vivisectionists praised the arguments of positive scientific knowledge and the utilitarianism that benefited human beings. Some also claimed that animals subjected to vivisection were a small fraction of all the animals killed or exploited for less noble reasons, highlighting the inconsistencies of those who claimed to be antivivisectionists and then ate meat or went hunting. Even Darwin refused to support the antivivisectionist cause, saying that it was unfair to demonize vivisectionists and not to blame those who shot birds for pleasure (Bianchi 2013). Others believed that pain could be inflicted to a greater or lesser extent, based on the scale of proximity to the human being of the different types of living beings (as theorized by the Italian Paolo Mantegazza, in his *La fisiologia del dolore*, 1880) (Pireddu 2014).

Complaints against vivisection were raised also in Italy, in Florence, by members of the local community and the English colony living there, condemning the experiments carried out by Moritz Schiff and Alexander Herzen at the newly founded Istituto di Studi Superiori (French 2019). Several articles were published in local newspapers as early as 1863. The first Italian society for the protection of animals[2] was created in Turin in 1871 by Giuseppe Garibaldi, "the hero of the Italian Risorgimento," together with Anna Winter and Timoteo Riboli, while enthusiastic physiologists like Paolo Mantegazza accused "England's bigots" of semi-religious hysteria (Pireddu 2014). In 1873 Schiff was taken to court. Herzen involved himself in the controversy with a booklet titled *Gli animali martiri, i loro protettori e la fisiologia* (1874), in which he wrote: "The martyred animals of physiological experiments are few compared to the martyred animals of other human activities (sustenance, comfort, luxury, 'gluttony', ignorance, whimsy, ferocity and vainglory, fun)."

In London, in 1875, the feminist, social reformer, and animal activist Frances Power Cobbe, with Anthony Ashley-Cooper, George Hoggan, and the Archbishop of York, William Thomson, formed the Society for the Protection of Animals Liable to Vivisection, better known as the Victoria Street Society (VSS). Among the members (and later both with the role of vice-president), were the poets Alfred Tennyson and Robert Browning. Then the Society split, and other antivivisection societies founded, so that by the end of the decade, there were at least a dozen of them only in Great Britain (Farmer 1996). Legislation to regulate vivisection was introduced by Parliament in 1876. It was not considered adequate by those opposing the practice, because animals could still be used, but they had to be anesthetized unless anesthesia interfered with the purpose of the experiment.

A heated debate on vivisection was triggered, to which some of the greatest scientific and literary figures of the time contributed. The matter came up in courtrooms and even caused demonstrations in public squares. A large number of texts on the subject circulated. As early as 1866, Charles Dickens had published an article titled *Inhumane Humanity* about vivisection in his journal, *All Year Round*, in which he wrote that "Man may be justified—though I doubt it—in torturing the beasts, that he himself may escape pain; but he certainly has no right to gratify an idle and purposeless curiosity through the practice of cruelty" (Dickens 1866, 240). In 1875, Lewis Carroll sent a letter to the *Pall Mall Gazette*, with the title *Vivisection as a Sign of the Times.* Here he stressed that science, demonstrating the closeness between humans and animals, required a renewed examination of human responsibilities towards nonhumans; in the same year, he also wrote an essay on the same subject, titled *Some Popular Fallacies about Vivisection.* In 1875 a former vivisector, George Hoggan, wrote a letter to the *Morning Post* denouncing the abuses committed in laboratories. The letter was later reprinted in *The Spectator* and sparked weeks of debate in the "Letters to the Editor" section (Farmer 1996). In 1885, John Ruskin resigned his Chair at Oxford, following a vote that permitted vivisection to be practiced in that University.

In England, the conflict reached its peak with the affair of the Brown Dog, a poor animal subjected to several demonstrations in the University of London, even if the 1876 Cruelty to Animals Act forbade the use of the same animal in more than one experiment. Two Swedish students, who had infiltrated the laboratory, exposed the case in a book entitled *The Shambles of Science* (Lind-af-Hageby and Schartau 1903). This led to a string of court cases and demonstrations. A monument dedicated to the Brown Dog was inaugurated in 1906 with a great event. George Bernard Shaw also participated as a speaker. Then pro-vivisection riots followed, and the statue was eventually removed in 1910 (a new statue was erected in 1985).

The debate echoed on the other side of the Atlantic Ocean, where the conflict was lively. In 1866 Henry Bergh had founded the American Society for the Prevention of Cruelty to Animals (ASPCA), while the American Anti-Vivisection Society (AAVS) was created in 1883. Even more so than in England, the antivivisectionist movement in the United States was marked by a large female presence and partly shared with feminism the fight against objectification, body control, and male domination.

Popular culture gave a vital contribution to the debate; in fact, literature was "mobilized" in this regard (Li 2006). Dogs played a leading role. Those in favor of vivisection tended to represent (and use) dogs as mere research tools. Antivivisectionists sought instead to stimulate the public's sensibility, highlighting not only that dogs were capable of suffering, but also that they had consciousness and sensitivity. In 1872, Frances Power Cobbe wrote two articles, *The Consciousness of Dogs* and *Dogs Whom I Have Met*, where both the intellectual and emotional capacities of the dogs were described, as well as their individuality: "There are few things more irritating to one who consistently honors dogs, than to hear superficial and indiscriminate people talk of those animals as if they were all alike in their mental and moral qualities, and only differed from each other by being white or black, rough-haired or silkycoated" (*Dogs Whom I Have Met*, 1872, 662).

Darwin also believed that dogs had a conscious life: "But can we feel sure that an old dog with an excellent memory and some power of imagination, as shown by his dreams, never reflects on his past pleasures in the chase? And this would be a form of self-consciousness" (1871, 83, ed. 1889). For this reason, after reading *The Consciousness of Dogs*, he sent a letter to Frances Power Cobbe (reprinted in Hill and Deegan 2016, 33–34), where, although criticizing some of her more daring statements, he wrote: "I agree with you on most points."

There was also a proliferation of short stories, poems, and novels. New forms of narrative were experimented with (Mayer 2009) to highlight the emotional subjectivity of animals, and to offer "a word for the dumb" (as suggested by the title of a poem by Christina Rossetti, 1896). In particular, several authors explored the path of "animal autobiography," a narrative strategy destined to achieve enormous success at the turn of the century with novels such as *Black Beauty: The Autobiography of a Horse* (1877 Anne Sewell) and *Beautiful Joe*. The experiment of narrating "beyond the human" (Herman 2016a, 1) represents an attempt to show solidarity with the way that nonhumans live in and see the world, and was "designed to promote more humane treatment of companion animals," even if "human frames of reference preponderate" (Herman 2016b, 11).

In 1867, Frances Power Cobbe published *The Confessions of a Lost Dog*, a novel narrated as the "canine autobiography" of a family dog, who was lost and then rescued. Ouida followed with her canine autobiography *Puck* (1870). Other canine autobiographies, often marked by extreme anthropomorphism, were published in their wake: *Sable and White: The Autobiography of a Show Collie*, by William Gordon Stables (1893); *Three of Us. Barney. Cossack. Rex,* by Izora C. Chandler (1895); *Hector, My Dog. An Autobiography*, by Egerton Ryerson Young (1905); and Mark Twain's short story *A Dog's Tale* (1903). Some of them were directly related to vivisection.

In North America, Mark Twain's story had a great effect. Like Tray, Twain's dog (named Aileen Mavourneen) saves a child, in this case from the flames. However, the gratitude of her owner, a scientist, cannot prevent Aileen's beloved puppy from being sacrificed for a vivisection experiment. Albeit similar in plot to Browning's poem, the different narrative format allows Twain, who writes from Aileen's point of view, to add elements to the analysis of the relationship between human beings and dogs. The distance between the good disposition of the dog, who accepts reproaches even if she does not understand the reason for them, and the authoritarian attitude of the human being is highlighted. This leads to the story's central point. Aileen does not understand what happened to her puppy, which is buried in the garden after the experiment. She hopes that he has been "planted" like a tree and waits for him to grow back.

Other works follow a more traditional form, like *Heart and Science: A Story of the Present Time*, by Wilkie Collins (1883), and *Trixy*. The latter work is not a canine autobiography, it has a more complex plot, including several characters, a court case, and even a love story. In the book, Elizabeth Stuart Phelps "engages the nuances of animal welfare campaigns during her own time, while presciently envisioning affiliation across the species divide as the basis for human compassion for animals" (VanDette 2019). Trixy is the trained poodle of a little orphan called Dan, who earns a little money thanks to her performances. Besides being his financial support, the poodle is also his emotional point of reference. Like Trixy, Miriam's beloved cocker spaniel Caro is also stolen. Miriam is Dan's landlady and, having recovered the two dogs, engages a lawsuit to prove that they are owned dogs, which she wins. The physiologist who performed his experiments on Caro is none other than Miriam's suitor Dr. Steele, a once gentle and sensitive doctor, but now heartless. Of course, he is unsuccessful as a suitor and dies alone amid the terrible memories of all the dogs he has dissected in his life. The novel touches on some specific points of the vivisection discussion, from a strongly anthropocentric perspective, such as the fear that doctors who practice vivisection become insensitive human beings or that the two dogs are precious pets owned by someone, not as sentient beings. However, the intensity of the connection between Dan and Trixy, and between Miriam and Caro, highlights the construction of interspecies families (Walker 2013) and the kinship between the two companion species reaches such an intensity as to prefigure a form of "early posthumanism" (Harde 2015).

Other important exponents of the culture of the time expressed themselves on the subject, some in favor of vivisection, some against. One remark made as a criticism towards many exponents of the movement was the lack of consistency. George Bernard Shaw, a long-standing opponent of vivisection, also pointed this out in the preface of his play *The Doctor's Dilemma* (1911, XLV):

"On one occasion I was invited to speak at a large Anti-Vivisection meeting in the Queen's Hall in London. I found myself on the platform with fox hunters, tame stag hunters, men and women whose calendar was divided, not by pay days and quarter days, but by seasons for killing animals for sport: the fox, the hare, the otter, the partridge and the rest having each its appointed date for slaughter. The ladies among us wore hats and cloaks and head-dresses obtained by wholesale massacres, ruthless trappings, callous extermination of our fellow creatures. We insisted on our butchers supplying us with white veal and were large and constant consumers of pate de foie gras; both comestibles being obtained by revolting methods. We sent our sons to public schools where indecent flogging is a recognized method of taming the young human animal. Yet we were all in hysterics of indignation at the cruelties of the vivisectors. These, if any were present, must have smiled sardonically at such inhuman humanitarians, whose daily habits and fashionable amusements cause more suffering in England in a week than all the vivisectors of Europe do in a year. I made a very effective speech, not exclusively against vivisection, but against cruelty…" (*The Doctor's Dilemma*, 1911).

Shaw thus condenses, in a few lines, the need for an ethical approach not only to vivisection but also to what one eats and what one wears (Preece 2009). Moreover, with the hint at the "domestication of the young human animal," he places himself firmly on the side of posthuman thought. H.G. Wells (1828) similarly accuses the antivivisectionists of inconsistency. He affirms that the pain inflicted in the scientific experiments is "an amount of suffering infinitesimal compared with the gross aggregate of pain inflicted day by day upon sentient creatures by mankind." Though, on this basis, he declares himself in favor of vivisection.

8.4 "Like a lady's ringlets brown": exploring the dog's *Umwelt* in *Flush*

Flush is the main character of a novel, today a little underrated, by Virginia Woolf (*Flush: a Biography*, 1933). The story is told from the point of view of the dog. The search for a narrative method capable of probing "not only potential heterogeneities but also potential areas of commonality in the structure of experience across the species boundary" (Herman 2016b, 54) had been successfully explored before, with the "canine autobiography." Woolf chooses a different strategy and introduces an extradiegetic narrator, the omniscient "biographer." Without pretending that the dog can talk, relevant information and comments are taken from the dog's owner's letters, introduced in the narrative by Woolf in a diegetic way. Still, not only is the story narrated from the dog's point of view, but the setting is seen (or better, smelled) from the dog's perspective (so, it becomes a "smellscape").

Unlike other literary dogs, Flush is a dog that really existed, having belonged to the poet Elizabeth Barrett. The quotation "Like a lady's ringlets brown" is from a poem that she dedicated to him. The novel is not as adventurous as London's *The Call of the Wild* or Phelps' *Trixy*, but it involves the "stolen-pet plot" (Walker 2013). Flush was indeed stolen, like many others in the second half of the nineteenth century, for a ransom (Howell 2000). Elizabeth Barrett was already known as a poet but still living with her parents, where she spent most of her time in her bedroom, due to her poor health. She had to oppose her father to pay the ransom and leave her parents' home, secretly and without their consent, to get the precious pet back.

Woolf builds Elizabeth's and Flush's stories along a parallel path. In the first part of the novel, they are both "captive" in Elizabeth's spacious family home in Wimpole Street. She is captive because of her poor health and paternal imposition; Flush because he is with Elizabeth. Out of love for her, he does not even have too much nostalgia for running free in the meadows. In London, Flush is partly protected by the Victorian rule of the chain, but at the same time the chain prevents him from moving freely. The turning point of the story is the kidnapping of Flush when he is not on a leash. To get her dog back, Elizabeth rebels, leaves the house, and goes in search of the thief herself. The success of this first insurrection opens the way to a much bigger rebellion: against her father's wishes, Elizabeth wants to marry the poet Robert Browning and to do so she runs away from home. The second part of the novel is a story of freedom for both Elizabeth and Flush. Elizabeth, now Mrs. Browning, moves to Italy, first to Pisa, and then to Florence, where she also has a baby. Flush discovers that in Pisa and Florence the rule of the chain is not applied. Thanks to his new freedom, he gets intoxicated by the smells and his unrestricted exploration of the city. In the final chapter, Elizabeth is getting older, but so is Flush. He dies, in silence, under her very eyes.

In Woolf's time, *Flush* was a successful novel, but it has since been judged a minor work (Johnson 2013). It has at most received consideration for the use of a dog as a metaphor for a woman, with Flush's confinement depicting the treatment of women in Victorian times. In this critical approach, with a strong anthropocentric bias, the dog has no relevance as an animal, merely as a metaphor (Smith 2002). However, other studies (Smith 2002; Johnson 2013; Feuerstein 2013) highlight the novel's value within a posthuman theoretical framework. As remarked by Feuerstein (2013, 32), "though Flush's canine epistemology allows us to move beyond observable aspects of Victorian patriarchy, this does not reduce Flush to the role of symbol or 'literary device,' as the novel's engagement with patriarchy simultaneously portrays the troublesome power relationships associated with pet keeping." Indeed, the two figures of Elizabeth and Flush are similar but different in their subordination, close but separate: "Between them lay the widest gulf that can separate one being from another. She spoke. He was dumb. She was woman; he was dog" (*Flush: a Biography,* 1933).

In the narrative, they undergo a parallel journey. The lack of freedom of one reflects the lack of freedom of the other, but not only in a metaphorical sense. The constant reference to the chain, and the contrast with the freedom of Italian dogs, speak directly about the canine condition. The division between the canine and human world in *Flush* remains open. At the same time, Elizabeth and Flush are "companion species" (Haraway 2003) not only for the content of the narrative but also for its construction. The extradiegetic narrator uses Flush's point of view as a reference point. Woolf may try to depict the world, using modes of perception of physical reality that are different from those usually employed for the description of the spatial context. In this way, the canine protagonist is endowed with a non-anthropomorphic perspective, demonstrating that Woolf "...ever attuned to the fluidities, volatilities, and disruptions that are a part of human relationships, was also sensitive to the ebbs and flows of the patterns of interactions among nonhuman—and human—animals" (Dubino 2012, 156).

Obviously, "attempts to articulate the inner lives of animals take place in unequal power relationships where humans ultimately have the power to describe and articulate animal subjectivity" (Feuerstein 2015, 151), as even the most imaginative authors cannot get out of their subjectivity. Animals are both producers and holders of knowledge, acquired through specific sensory apparatuses, even if space is usually qualified only by the human dimension (Raffestin 1986, 133–134). Specifically, as Raffestin points out, Jakob Johann von Uexküll demonstrated how the subject creates space and time (von Uexküll (1934)). For the German biologist, every animal has its own "phenomenal world" (*Umwelt*). Penetrating an animal's *Umwelt* — that is entering the subjective space and time of animals — is impossible. "There simply is no way to truly know the consciousness of a dog" (Johnson 2013, 35). Thus, to pose oneself from the point of view of a nonhuman animal requires an excellent imagination. Woolf herself confesses a human's "inadequacy" to describe the world through smells; nevertheless, she carefully points out that for Flush, the landscape is a landscape of odors.

Attempting to filter the world through a dog's point of view might open a perspective that somehow allows us to look at our world and the lives of other animals through different lenses. Indeed, "the extreme phenomenological bracketing required to 'understand' what it might be like to 'think like a dog' helps us to reframe our conceptions of 'the other' in dramatic and instructive ways" (Hill and Deegan 2016, 1). In the case of Flush, "canine epistemology, which functions primarily by way of scent, challenges the empirical belief in the authority of vision and the ability to know and understand simply by looking" (Feuerstein 2013, 32).

In Woolf's descriptions, the reader is introduced to Flush's spatial experiences through smells: the smell of cologne when the dog enters the room of Elizabeth for the first time, the smell of earth and grass when he goes to the park, the change of smell when, finally free from the "captivity" of Wimpole Street, they arrive in Italy. "There was none of that heady confusion of smells that made a walk down Wimpole Street or Oxford Street so distracting. On the other hand, the strange new smells that came from sharp stone corners, from dry yellow walls, were extraordinarily pungent and queer… ." (*Flush: a Biography,* 1933).

The need to embrace a change of perspective returns in another text by Virginia Woolf, *Gypsy, the Mongrel* (1940), where the author formulates the question directly: "Do dogs see what we see or is it something different?"

8.5 "Friends in a friendship closer than brotherhood": Ouida and *A Dog of Flanders* in posthuman perspective

The same question, or at least the attempt to answer it, is present in Ouida's novel *A Dog of Flanders* (1872). Written by an English novelist, famous while she was alive but since forgotten (King 2015), the book was a great success when it came out. Then it and its author fell into oblivion in her homeland, though it became a classic of children's literature in Japan. It inspired a cult animé capable of attracting Japanese tourists to the city where it is set, Antwerp, and of stimulating the transformation of the urban iconography to satisfy those tourists.

A Dog of Flanders is about a friendship between a dog called Patrasche and a boy named Nello. Nello is a little orphan, living with his grandfather, who is very poor and very old. To scrape together a few coins, the grandfather delivers milk to the city, carrying it with an old cart from the village. Nello goes with him. One day, they find a big dog abandoned half-dead by the side of the road, after being brutally beaten. They decide to take him home and try to save his life. After a few days, the dog recovers and, being a dog trained to pull a cart (as was the custom in Flanders at the time), he shows his gratitude pulling the cart with the milk to the city. In this way, Patrasche becomes the "bread-winner and minister" for the grandfather and the child. He is also "their only friend and comforter."

As the grandfather gets older, the child grows up. He cannot go to school because he is penniless, but learns to draw by himself, and he draws very well. Apart from his dog, he has no friends except for a little girl named Alois, the mill-owner's daughter. For some years, the lives of Nello and Patrasche flow smoothly, despite their poverty. Growing up, Nello is too poor to be Alois' friend: her father, worried that their friendship will turn into love, forbids the two children from continuing to see one another. Nello dreams of seeing the great paintings by Rubens in the Cathedral but he is too poor also for this; to see them, he would have to pay for a ticket, and he has no money. The only worry that Patrasche suffers is to see Nello worried.

At a certain point, things fall apart: the grandfather dies, Nello is wrongfully accused of arson and loses his customers and the solidarity of his fellow villagers. He has no money for rent and so gets evicted from the hut he lives in. His last hopes lie in a drawing contest, which he expects to win, but he loses. On Christmas Eve, having left Patrasche in the care of Alois' family, he seeks refuge in the cathedral in Antwerp, where the Rubens paintings are. Patrasche runs away from Alois to find his friend. The next morning, they are found dead from hunger and the cold but in each other's embrace in front of Rubens' triptych. Together, they will be buried in the same grave.

According to some critics, *A Dog of Flanders* is a children's story (Pollock 2005). Others consider it on several levels (King 2016) and as a text on animal rights (Pollock 2005). From this point of view, the extraordinary friendship between the dog and the child seems to be a bond that transcends the differences between human and nonhuman species, overcoming the animal/human distinction from a posthuman perspective. And, as such, the text appears to correspond to the interpretative approach offered by Donna Haraway (2003).

At the beginning of her writing career, Ouida had become famous for writing "sensation novels," such as *Held in Bondage* (1863), *Strathmore* (1865) and *Under Two Flags* (1867) (Jordan 2011), which feature a mixture of adventure, intrigue, and romance that was well suited to Victorian popular literary tastes (Gilbert 1997). The characters were somewhat unlikely aristocrats, often sexually promiscuous (a trait that was also typical of her female characters, as long as they were villains). Her later novels spoke instead of the destitute (Pollock 2005) and were tinged with a touch of political criticism (Gilbert 1997). Several of her female characters were unconventional, refusing gender roles, questioning marriage, and living a life of activity and self-determination (Jordan 2011). *A Dog of Flanders* belongs to this second phase. The characters are humble, they work to subsist until their situation changes, and when they lose the opportunity to work, they starve to death. Nello is a boy who cannot go to school.

The dog comes from "a race which had toiled hard and cruelly from sire to son in Flanders many a century—slaves of slaves, dogs of the people, beasts of the shafts and the harness, creatures that lived straining their sinews in the gall of the cart, and died breaking their hearts on the flints of the streets" (*A Dog of Flanders,* 1872).

Alois' mother, the only female character apart from little Alois herself, is the only person who sees that love can bring happiness, even without money. She is also the only one who always takes Nello's side. Of course, her husband does not listen to her; instead, he says: "You are a woman, and therefore a fool" and forbids the children from seeing each other. Even if the subordinate condition of women or of the poor — exploited and deprived of the possibility of putting their talents to good use — is not the central theme of the novel, the text opens a fissure in the capitalist and patriarchal discourse. From a literary point of view, Ouida is not interested in theories (Pollock 2005). The short novel is not a canine autobiography (like the earlier *Puck*, by the same author) and does not investigate the dog's *Umwelt*, as Virginia Woolf does in *Flush*, but still tries to offer the dog's point of view. What makes the story noteworthy is that no point of view takes precedence over the other, neither the human one nor the canine. The narrator is external, extradiegetic, but follows, in parallel, both Nello's and Patrasche's perspectives.

From the *incipit* to the end, what matters is the bond that connects Nello and Patrasche. We know that they are individuals who have chosen to be friends. At the same time, the author reminds us that they are different. They are the same age "by length of years," but one is still young, and the other is old. Nello thinks and articulates his dreams. The dog cannot understand everything, but he gives his explanation. From his point of view, Antwerp's churches are "great, sad piles of stone," while the small church of the village is a "small, tumbledown, gray pile opposite the red windmill." Rubens, the great Flemish painter much admired by Nello, must have been a good person since Patrasche knows that he "had loved dogs or he had never painted them with such exquisite fidelity; and men who loved dogs were [...] always pitiful" (*A Dog of Flanders,* 1872).

The tale of Nello and Patrasche was translated into Japanese for the first time in 1908 by Shiken Hidaka. It was subsequently rendered into many media, including two animé TV series, in 1975 and 1992, an animated film (1997), and even a video game. Nowadays, in Japan, "Nello and Patrasche" "is a phrase representing a bond never to be untied ..." (Ikawa 2018, 1).

Thanks to this success, the story lures many visitors from Japan to Antwerp (Van Broeck and De Gruyter 2004; Geraghty 2019). There are no material references that can satisfy their curiosity in Hoboken's village, where Nello's house is believed to have stood. So a small statue of the boy and the dog was erected there in January 1985. In 2016, just outside of the Cathedral of Our Lady, a marble statue by Batist Vermeulen (Tist) was erected of Nello and Patrasche huddled together under a blanket of cobblestones. According to visitors' comments on Tripadvisor, it makes you want to go home and hug your dog.[3] Hence, even if in a rather tortuous and transmedial way, via the novel, the animé, and a marble statue in Antwerp's public space, Ouida's message still remains powerful today.

8.6 "Jara is my friend": antispeciesism (and environmental justice) in *Animal's People*

Animal's People (Indra Sinha, 2007) is a novel set in the fictional location of Khaufpur ("pur" is a suffix that means "village," while "khauf" in Urdu means "fear"). So, Khaufpur is the "place of fear." Khaufpur stands for Bhopal, the Indian city where "the worst industrial crisis in history" (Kim 2014) occurred on the night of December 2–3, 1984. The disaster, known as the "Bhopal Gas Tragedy," was caused by the leakage of about 45 tons of methyl isocyanate, MIC, from a pesticide plant owned by the Indian subsidiary of the US corporation Union Carbide. About 500,000 people were exposed to the gas. The number of victims varies between the 3,000 deaths officially declared at the time of the disasters and the 15,000 subsequently estimated. Hundreds of thousands of other people then suffered from various forms of infirmity due to gas exposure, in what Rob Nixon described as a deadly form of "slow violence" (Nixon 2011). According to a survey conducted by the Indian Medical Research Council, the vast majority of people affected by the gas were of very low social and economic status (Faras 2020).

Like *The Road* (2006, Cormac McCarthy), and *The Walking Dead* (2010-on, AMC), *Animal's People* does not describe the disaster. The author focuses on "the less obviously eventful aftermath, the slow violence that, by the novel's end, comes to be recognized as the event itself, a violence that has yet to run its course" (Nixon 2011, 44). The story develops around the efforts of a group of activists to get justice about 20 years after the disaster. It is narrated in the first person by Animal, one of the survivors still suffering the consequences of the explosion, in a language that blends English, French, Hindi, and many words transliterating English terms in a very personal way (the disaster is called "the Apokalis," and Union Carbide is "the kampani"). Animal is a 19-year-old boy who was born during the days of the catastrophe. At birth, he was a child like all the others. He had a name, which nowadays he does not remember. Affected by the gas, he is now called "Animal" because, instead of walking erect, he moves on all fours. He is one of "the people of the Apokalis" (one of the victims of the disaster).

The novel raises many issues. The first is environmental justice, or rather injustice (Faras 2020), in a transcalar perspective (Taylor 2013), from the cellular (Animal's body) to transnational levels (the connections between the Kampani and the people of Khaufpur). In this way, "The novel gives focus to three of the defining characteristics of the current neoliberal order: first, the widening chasm—within and between nations—that separates the megarich from the destitute; second, the attendant burden of unsustainable ecological degradation that impacts the health and livelihood of the poor most directly; and third, the way powerful transnational corporations exploit [...] the lopsided universe of deregulation, whereby laws and loopholes are selectively applied in a marketplace a lot freer for some societies and classes than for others" (Nixon 2011, 46).

If decisions are taken elsewhere, local populations are deterritorialized, i.e., they have no power over what happens in their territorial context (Raffestin 1984). For this reason, writes Rob Nixon (2011, 65), Khaufpur works like a synecdoche, about all the "poisoned communities spread out across the global South."

The text, framed as a translation from a transcription of the tapes recorded by Animal, is presented as a narration made at first to an Australian journalist (a "jarnalis" from "Ostrali"), then directly to the readers (that Animal calls "eyes"). Animal accuses the journalist of being attracted to his story because he is "drawn by the smell of blood." This possible meta-reading opens the second issue: a certain strand of contemporary literature "that mines sites of political violence for stories of suffering and positions the reader as a concerned humanitarian" (Rickel 2012, 87). Moreover, thanks to its narrative structure, it "tells the story of a young subaltern individual fighting for his right to ignore common ways of telling the story of this tragedy" (Bartosch 2012, 11). Written "as if" it were a subaltern speaking in a long first-person monologue, the text is presented as a voice from below. In this way, it engages the debate urged by Spivak (1988) on the possibility for subalterns to speak to a dominant audience (Rath 2013, 162). The real author of the text is Indra Sinha, who is not a subaltern, but a successful writer; so, as in all pseudo autobiographies, he can only use his imagination to place himself in the position of the protagonist and adapt to his *Umwelt*.

Animal has many friends, including some community residents, an old French nun, and, above all, the dog Jara. "Jara is my friend": in this way, Animal introduces her. The relationship between Animal and Jara places emphasis on the posthumanist understanding. Animal and Jara met when, both hungry and helpless, both on all fours, they contended for leftovers from restaurants. Later they became partners and learned to give, always outside restaurants, a small performance: she pretends to fall dead, while he begs. They share the same point of view because they do not see the world of humans, which is viewed at a human's eye-level; they see below the belt. Unlike Virginia Woolf, Sinha does not present the world through the olfactory perceptions of a dog. Instead, he perceives the world at the level of a dog's eyes. As a dog, Animal "gains an intimate understanding of the geography of Khaufpur" (Kim 2014), since: "from a height of eighteen inches you get to know a place pretty well, every crack in the road, every stone, every dropped, not-picked-up coin" (*Animal's People*, 2007).[4]

The relationship between humans and nonhuman animals is one of the central themes. Animal introduces himself with the phrase: "I used to be human once." A similar statement is variously repeated throughout the novel. Following Mukherjee (2011, 221), "Animal's proclamation of his nonhuman identity gives voice to a scandal that lurks behind the tragedy of Bhopal — if there are those who, by dint of their underprivileged location in the hierarchy of the "new world order," cannot access the minimum of the rights and privileges that are said to define humanity, what can they be called? It also lends voice to a different set of questions — what happens when the rights and privileges of humans are achieved explicitly at the expense of the sufferings of the majority of nonhuman beings, and how can we imagine an alternative practice"?

Not being human, however, for Animal does not mean being less than human. It is a choice. It means to be on the side of those who have no rights (and for this reason, they have "the invincible, undefeatable power of zero"). Thus, "Animal is human, but he struggles to figure out a posthumanist way of understanding his existence and his relationship to others" (Rickel 2012, 91). This issue opens a perspective that questions anthropocentrism frontally, as it resonates in the discussion about Paradise: "'Farouq says if I want to end up in paradise I'll have to turn human.' 'Why so, moosh?' 'Paradise is for humans, not for animals.' 'What harm do animals do?' 'Not a question of harm. Do you expect that every ant that gets crushed under a villager's horny heel goes to paradise?' 'Don't see why not. If they have flowers and birds in paradise why not ants? Isn't there room?'" (*Animal's People*, 2007).[5]

A few more elements help to unlock the author's anti-anthropocentric approach. First of all, there is Animal's bond with Jara. Then there is Somraj, the former singer, who has lost his ability to sing as a consequence of the disaster and now teaches music, not only to children, but also to lizards, and recognizes musical notes in the singing of frogs. Then, more related to the use of language, there is the frequent use of similes with animals, dotting Animal's narration. With this constant reference to the animal world, a sense of common space is created not only between Animal and the nonhuman animal world, but also between the nonhuman animal world and all the members of the community, "Animal's people." The setting offers some additional elements. In the area of the factory, after the gas leak, "A forest is growing, tall grasses, bushes, trees, creepers that shot sprays of flowers like fireworks…throughout this place a silent war is being. Mother Nature's trying to take back the land…at such moment the forest is beautiful, you forget it's poisoned and haunted" (*Animal's People*, 2007).[6]

Like many other protagonists of dystopian or post-apocalyptic narratives, Animal finds refuge in the forest. The forest of Bhopal, however, is not a "green place" and does not offer a way out through classic geographical binarism. It is poisonous, though lush. It thus reminds us that the world is one; there is no Planet B. In the forest, Animal gets lost and, starving, he believes he is dying. Jara, followed by his human friends, finds him. After being rescued, Animal decides that the surgery which could make him walk erect again is no longer his goal. He does not need it. He is Animal, and his "people" are made up of humans and nonhumans. He leaves the readers with a provocative reminder: "Tomorrow there will be more of us." More desperate people, or more "animal's people," sharing a sense of collective belonging and community between humans and nonhuman animals?

Notes

1 The Kennel Club was founded in 1873; it therefore existed at the time the novel was written, but not at the time of the life of Elizabeth Barrett, who died in 1861, or of Flush. A Spaniel Club was only founded in the UK in 1885.
2 "Società Protettrice degli Animali contro i mali trattamenti che subiscono dai guardiani e dai conducenti," Society for animal protection against the evil treatment they suffer from their guardians and drivers, later Ente Nazionale Protezione Animali, ENPA.
3 www.tripadvisor.com/ShowUserReviews-g188636-d12645595-r731029874-Nello_Patrasche_Statue-Antwerp_Antwerp_Province.html — consulted on April 4, 2020.
4 The excerpt, from the novel *Animal's People*, Indra Sinha, Simon & Schuster, is quoted with the permission of the publishing house.
5 The excerpt, from the novel *Animal's People*, Indra Sinha, Simon & Schuster, is quoted with the permission of the publishing house.
6 The excerpt, from the novel *Animal's People*, Indra Sinha, Simon & Schuster, is quoted with the permission of the publishing house.

Bibliography

Armbruster, Karla (2002). Good Dog: The Stories We Tell about Our Canine Companions and What They Mean for Humans and Other Animals. *Papers on Language & Literature* 38(4), 351–376.

Barth, Fredrik (ed.) (1969). *Ethnic Groups and Boundaries, The Social Organization of Culture Difference* (pp. 9–16). Little, Brown and Company.

Bartosch, Roman (2012). The Postcolonial Picaro in Indra Sinha's Animal's People – Becoming Posthuman through Animal's Eyes. *Ecozon@* 3(1), 10–19. https://doi.org/10.37536/ECOZONA.2012.3.1.446

Bates, A.W.H. (2017). *Anti-Vivisection and the Profession of Medicine in Britain*. Palgrave Macmillan.

Bianchi, Bruna (2013). "Come i secchi nel pozzo." Scienza ed etica negli scritti contro la vivisezione delle femministe britanniche (1870–1910). *DEP. Deportate, esuli, profughe. Rivista telematica di studi sulla memoria femminile* 23, 4–31.

Carroll, Lewis (1875). Some Popular Fallacies about Vivisection. *The Fortnightly Review* 23, 847–854.

Darwin, Charles (1871). *The Descent of Man and Selection in Relation to Sex*. John Murray.

Dickens, Charles (1866). *Inhumane Humanity. All the Year Round* XV(17), 238–240.

Dubino, Jeanne (2012). The Bispecies Environment, Coevolution, and Flush. In Ryan, Derek, & Bolaki, Stella (eds.), *Contradictory Woolf: Selected Papers from the Twenty-first Annual International Conference on Virginia Woolf* (pp. 150–157). Clemson University Digital Press.

Faras, Sabiha Asif (2020). Environmental Injustice in Indra Sinha's Animal's People. *Our Heritage* 68(32), 185–190.

Farmer, Steve (1996). Introduction, in Collins, Wilkie, *Heart and Science: A Story of the Present Time* (pp. 7–27). Broadview.

Feuerstein, Anna (2013). What Does Power Smell Like? Canine Epistemology and the Politics of the Pet in Virginia Woolf's Flush. *Virginia Woolf Miscellany* 84, 32–34. https://virginiawoolfmiscellany.wordpress.com/virginia-woolf-miscellany-archive-issue-84-fall-2013-through-issue-92-fall-2017-winter-2018/

Feuerstein, Anna (2015). "I Promise to Protect Dumb Creatures": Pastoral Power and the Limits of Victorian Nonhuman Animal Protection. *Society & Animals* 23(2), 148–165. https://doi.org/10.1163/15685306-12341339

French, Richard D. (2019). *Antivivisection and Medical Science in Victorian Society*. Princeton University Press.

Fuller, Steve (2015). Evolution. In Ranisch, Robert, & Sorgner, Stefan Lorenz (eds.), *Post- and Transhumanism: An Introduction* (pp. 201–211). Peter Lang.

Geraghty, Lincoln (2019). Destination Antwerp! Fan Tourism and the Transcultural Heritage of A Dog of Flanders. *Humanities* 8(90), 1–11. https://doi.org/10.3390/h8020090.

Gilbert, Pamela K. (1997). *Disease, Desire, And The Body In Victorian Women's Popular Novels* Cambridge University Press.

Goldmark, Daniel, & McKnight, Utz (2008). Locating America: Revisiting Disney's Lady and the Tramp. *Social Identities: Journal for the Study of Race, Nation and Culture* 14(1), 101–120. https://doi.org/10.1080/13504630701848705

Haraway, Donna (2003). *The Companion Species Manifesto: Dogs, People, and Significant Otherness*, Prickly Paradigm Press (also in Haraway, Donna. Manifestly Haraway. University of Minnesota Press, 2016).

Harde, Roxanne (2015). "Doncher be too sure of that!": Children, Dogs, and Elizabeth Stuart Phelps's Early Posthumanism. *Bookbird: A Journal of International Children's Literature* 53(1), 10–23 https://doi.org/10.1353/bkb.2015.0016

Herman, David (2016a). Animal Autobiography; Or, Narration beyond the Human. *Humanities* 5(82). https://doi.org/10.3390/h5040082.

Herman, David (2016b). Building More-than-Human Worlds: Umwelt Modelling in Animal Narratives. In Gavins, Joanna, & Lahey, Ernestine (eds.), *World Building: Discourse in the Mind* (pp. 53–70). Bloomsbury Academic.

Herzen, Alessandro (1874). *Gli animali martiri, i loro protettori e la fisiologia*. Bettini.

Hill, Michael, & Deegan, Mary Jo (eds.) (2016). *Dogs and Society. Anglo-American Sociological Perspectives (1865–1934)*. Zea Books.

Hoggan, George (1875). Vivisection, *Morning Post* 2 February,

Howell, Philip (2000). Flush and the Banditti: Dog-Stealing in Victorian London. In Philo, Chris, & Wilbert, Chris (eds.), *Animal Spaces, Beastly Places. New Geographies of Human–Animal Relations*. (pp. 35–55). Routledge.

Howell, Peter (2012). Between the Muzzle and the Leash: Dog-walking, Discipline, and the Modern City. In Atkins, Peter (ed.), *Animal Cities: Beastly Urban Histories* (pp. 235–256). Routledge.

Ikawa, Eri (2018). 井 川 恵 理. ネロとパトラッシュ—「フランダースの犬」における二つのベルギーの融合 (Nello and Patrashe: Merging of Two Belgiums in "A Dog of Flanders"). *J1SLA* 10,1−12.

Johnson, Jamie (2013). Virginia Woolf's Flush: Decentering Human Subjectivity through the Nonhuman Animal Character. *Virginia Woolf Miscellany* 84, 34–36. https://virginiawoolfmiscellany.wordpress.com/virginia-woolf-miscellany-archive-issue-84-fall-2013-through-issue-92-fall-2017-winter-2018/

Jordan, Jane (2011). Ouida. In Gilbert, Pamela K. (ed.), *A Companion to Sensation Fiction* (pp. 230–231). Wiley-Blackwell.

Kean, Hilda (1998). *Animal Rights: Political and Social Change in Britain since 1800*. Reaktion Books.

Kean, Hilda (2003). An Exploration of the Sculptures of Greyfriars Bobby, Edinburgh, Scotland, and the Brown Dog, Battersea, South London, England. *Society & Animals* 11(4), 353–373. https://doi.org/10.1163/156853003322796082

Kim, Jina (2014). "People of the Apokalis": Spatial Disability and the Bhopal Disaster. *Disability Studies Quarterly* 34(3). http://dx.doi.org/10.18061/dsq.v34i3.3795

King, Andrew. (2015) Ouida (Marie Louise Ramé). In Felluga, Dino Franco, Gilbert, Pamela K., Hughes, Linda K ., & Malone, Katherine (eds.), *The Encyclopedia of Victorian Literature* (pp. 1–7). Wiley- Blackwell. [https://doi.org/10.1111/b.9781118405383.2015.x]

King, Andrew (2016). Impure Researches, or Literature, Marketing and Aesthesis. The Case of Ouida's 'Dog of Flanders'(1871 – today). *English Literature* 2(2), 359–382. http://dx.doi.org/10.14277/2420-823X/EL-2-2-15-10

Lee, Vernon (1882). Vivisection: An Evolutionist to Evolutionists. *Contemporary Review* 41, 788–811.

Li, Chien-hui (2006). Mobilizing Literature in the Animal Defense Movement in Britain, 1870–1918. *Concentric: Literary and Cultural Studies* 32(1), 27–55. www.concentric-literature.url.tw/issue_detail.php?issue_title=Animals

Lind-af-Hageby, Lizzy, & Schartau, Leisa Katherina (1903). *The Shambles of Science: Extracts from the Diary of Two Students of Physiology*. E. Bell.

Mantegazza, Paolo (1880). *La Fisiologia del Dolore*, Paggi.

Mayer, Jed (2009). Representing the Experimental Animal: Competing Voices in Victorian Culture. In McFarland, Sarah E., & Hediger, Ryan (eds.), *Animals and Agency: an Interdisciplinary Exploration* (pp. 181–205). Brill.

McHugh, Susan (2004) *Dog*. Reaktion Books.

Mills, Brett (2017). *Animals on Television: The Cultural Making of the Non-Human*. Springer.

Molloy, Claire (2011). *Popular Media and Animals*. Springer.

Morris, David Copland (1996). A Dog's Life: Anthropomorphism, Sentimentality, and Ideology in John Muir's *Stickeen*. *Western American Literature* 31(2), 139–157. https://doi.org/10.1353/wal.1996.0104

Morse, Deborah Denenholz, & Danahay, Martin A. (eds.) (2017) *Victorian Animal Dreams: Representations of Animals in Victorian Literature and Culture*. Routledge.

Mukherjee, Pablo (2011). Tomorrow There Will Be More of Us. In DeLoughrey, Elizabeth, & Handley, George B. (eds.), *Postcolonial Ecologies: Literatures of the Environment* (pp. 216–231). Oxford University Press.

Nixon, Rob (2011). *Slow Violence and the Environmentalism of the Poor*. Harvard University Press.

Onion, Rebecca (2009). Sled Dogs ofThe American North: On Masculinity, Whiteness, And Human Freedom. In McFarland, Sarah E., & Hediger, Ryan (eds.), *Animals and Agency: an Interdisciplinary Exploration* (pp. 129–155). Brill.

Perkins Gilman, Charlotte (1911), On Dogs. *The Forerunner* 2,180–82, 206–209.

Pireddu, Nicoletta (2014). Between Darwin and San Francesco: Zoographic Ambivalences in Mantegazza, Ouida, and Vernon Lee. *Gothic Studies* 16(1), 111–127. https://doi.org/10.7227/GS.16.1.9

Pollock, Mary Sanders (2005). Ouida's Rhetoric of Empathy: A Case Study in Victorian Anti-Vivisection Narrative. In Pollock, Mary, & Rainwater, Catherine (eds.), *Figuring Animals: Essays on Animal Images in Art, Literature, Philosophy and Popular Culture* (pp. 135–159). Palgrave Macmillan.

Power Cobbe, Frances (1872a). The Consciousness of Dogs. *Quarterly Review* 133, 419–451.

Power Cobbe, Frances (1872b). Dogs Whom I have Met. *Cornhill* 26, 662–678.

Preece, Rod (2009). *Sins of the Flesh: A History of Ethical Vegetarian Thought.* UBC Press.

Raffestin, Claude (1984). Territorializzazione, deterritorializzazione, riterritorializzazione e informazione. In Turco, Angelo (ed.), *Regione e regionalizzazione* (pp. 69–82). Franco Angeli.

Raffestin, Claude (1986). Perché "noi" non abbiamo letto Eric Dardel? In Eric Dardel. *L'uomo e la terra. Natura della realtà geografica* (pp. 129–143). Unicopli.

Ranisch, Robert, & Lorenz Sorgner, Stefan (2015). Introducing Post- and Transhumanism. In Ranisch, Robert, & Lorenz Sorgner, Stefan (eds.), *Post- and Transhumanism: An Introduction* (pp. 7–27). Peter Lang.

Rath, Brigitte (2013). "His words only?" Indra Sinha's Pseudotranslation Animal's People as Hallucinations of a Subaltern Voice. *AAA: Arbeiten aus Anglistik und Amerikanistik* 38(2), 161–183.

Reynier, Christine (2009). *Virginia Woolf's Ethics of the Short Story*. Springer. www.jstor.org/stable/43025856

Rickel, Jennifer (2012). "The Poor Remain": A Posthumanist Rethinking of Literary Humanitarianism in Indra Sinha's Animal's People. *ariel: A Review of International English Literature* 43(1), 87–108. https://journalhosting.ucalgary.ca/index.php/ariel/article/view/35064

Ryder, Richard D. (2006). Speciesism in the Laboratory. In Singer, Peter (ed.), *In Defense of Animals: The Second Wave* (pp. 87–103). Blackwell Publishing.

Shaw, George Bernard (1911). *The Doctor's Dilemma*. Brentano's.

Smith, Craig (2002). Across the Widest Gulf: Nonhuman Subjectivity in Virginia Woolf's *Flush*. *Twentieth Century Literature* 48(3), 348–361. https://doi.org/10.1215/0041462X-2002-4003

Spivak, Gayatri Chakravorty (1988). Practical Politics of the Open End. *Canadian Journal of Political and Social Theory* 12(1–2), 104–111.

Taylor, Jesse Oak (2013). Powers of Zero: Aggregation, Negation, and the Dimensions of Scale in Indra Sinha's Animal's People. *Literature and Medicine* 31(2), 177–198. https://doi.org/10.1353/lm.2013.0014

Tuan, Yi-Fu (1984). *Dominance & Affection: The Making of Pets.* Yale University Press.

VanDette, Emily E. (2019). Introduction. In Stuart Phelps, Elizabeth, *Trixy: A Novel.* Northwestern University Press.

Van Broeck, Anne Marie, De Gruyter, Ilse (2004). Japanese Tourists: Looking for a Dog and its Master in Flanders. In Robinson, Mike, Picard, David, & Culver-Dodds, William (eds.), *Tourism and Literature: Travel, Imagination and Myth. Proceedings of the Tourism Literature: Travel, Imagination and Myth International Research Conference, Harrogate United Kingdom, 22–26 July 2004* (pp. 1–17). Centre for Tourism and Cultural Change, Sheffield Hallam University.

Von Uexküll, Jakob (1934). Streifzüge durch die Umwelten von Tieren und Menschen: Ein Bilderbuch unsichtbarer Welten. J. Springer (2010, *A Foray into the Worlds of Animals and Humans: With a Theory of Meaning*. University of Minnesota Press).

Walker, Alyssa Chen (2013). Bringing the Laboratory Dog Home: Elizabeth Stuart Phelps and the Antivivisection Narrative. *Humanimalia: A Journal of Human/Animal Interface Studies* 4(2), 101–129. www.depauw.edu/humanimalia/issue%2008/walker.html

Wells, H.G. (1928). Popular Feeling and the Advancement of Science. Anti-Vivisection. In Wells, H.G., *The Way the World Is Going: Guesses and Forecasts of the Years Ahead* (pp. 221–230). Ernest Benn Limited.

9 Posthuman (dis)orders

Monsters, hybrids, metamorphosis

9.1 Fantastic beasts, monsters and non-"normate" bodies, from Homer to *Harry Potter*

What is a monster? According to the definition offered by an Italian dictionary,[1] a monster is a being with characteristics that are extraneous to the usual order of nature. By this definition, the monster is not "ordinary," in the sense that it does not belong to the order. Therefore, the monster creates disorder. Creating disorder means getting out of binary oppositions, classifications, and social hierarchies. Culture has the function of assigning positive, negative, or neutral values to phenomena and symbols. Therefore, the meaning of monsters fluctuates, going from wonder to fear to curiosity.

The term "monster," which comes from the Latin *monstrum*, that is portent, has a negative meaning in many contemporary idioms. In journalistic language, the monster is someone who has committed a nefarious act, while in popular parlance a monster is just a hideous person. If, nowadays, the idea of the monster as "not ordinary" is conceived in negative terms, in the past, it had much more fluid meanings. Monstrous and hybrid figures are part of the symbolic heritage of humanity "since the beginning of recorded civilization" (Ferrando 2019), but not all hybrids are monsters, in the contemporary negative sense. In Indian mythology, for instance, extra-ordinary entities are often benevolent. In ancient Egypt, divinities have a hybrid aspect, like Set, with the body of a man and the head of an animal, or the Sphinx, with the head of a woman and the body of a lion. In Greek and Latin culture, to be "out of the ordinary" may have different meanings. In both cases, the *extra ordinem* may be a bridge between the divine and the human world (Coppola 2018). In ancient Greece, the term closest to "monster" was *teras* (Τέρας), generally translated as "marvel," "divine sign," "prodigy," or even "monstrous birth" (Li Causi 2013). As Baglioni writes (2013, 17), whatever is Τέρας, even if it manifests itself, at least in Homeric poems, at the initiative of the divinity, does not necessarily transmit a message. If it does, it is not necessarily a negative message, nor does it manifest itself only as a result of infractions of the norms. In Greek myths, hybrid beings are frequent, formed by a combination of limbs belonging to different animal species, or human and bestial bodies, and such deviations from the natural order do not necessarily have a negative connotation. The minotaur, the fruit of an aberrant coupling between the Queen of Crete and a beautiful white bull, certainly has; not only does it have the head of a bull on the body of a man (but with hooves, cow fur, and tail), it also feeds on human flesh. Other hybrid beings belong to a different order, like Chiron, a wise and benevolent centaur, the master of great heroes like Ajax, Achilles, and Aeneas.

In the Roman context, the term *monstrum* appears in association with events unrelated to the ordinary course of events, or to indicate deformed or hybrid beings. In both cases, a monster is an *omen* of doom instigated by the gods (Baglioni 2013), and in Rome animals with three legs, lambs with two heads, or hermaphrodites were understood as the result of a "violation of the order" (Li Causi 2013, 60). Beyond the *terata* and *monstra*, which could suddenly burst into the daily space as a sign of the gods, there were "monstrous" beings from different spaces, such as the manticore, a being with a human face and feline body, who lived, depending on the author, in India or Ethiopia; or from a distant past such as Scylla and Charybdis (Li Causi 2013). In all cases, the monsters defined the "limits of the community" (Haraway 1985) in both a spatial and temporal sense.

For the Roman culture, Pliny the Elder's *Natural History* (77–78 CE) was an essential source of information about "monsters." More came with travel accounts. For instance, in his *Chorographia* (37–41 CE), the geographer Pomponius Mela speaks of Acephals, Macrocephals, vampires, and werewolves. In the Middle Ages, the Bible and the Nordic legends added their old myths to the classic tales, and popular fantasies were filled with hybrid creatures, dragons, mermaids, etc. The *Liber monstrorum de diversis generibus*, written by an anonymous author probably in the eighth century, is an extended classification of monsters and wonderful creatures, divided into three categories: *De monstris* (almost human beings), *De belvis* (animals), and *De serpentis* (snakes). Dante's *Divine Comedy* is full of references to classical culture, including Cerberus, the Greek three-headed dog. Monsters and hybrids are also included visually, becoming a recurrent subject of medieval art (Sebenico 2005). Even angels, traditionally portrayed in Christian culture as asexual beings of extraordinarily harmonious ("angelic") forms, but endowed with large feathered wings, can be considered hybrids. Consequently, in the Middle Ages, "monsters" were ambiguous; they may be linked to the devil, but they also appear in Christian iconography. It is not certain that they are malicious; on the contrary, some extra-ordinary creatures may have a different appearance and composition; some monsters, indeed, are beautiful (Sebenico 2005).

The religious, or magical, dimension of the extra-ordinary creature was still predominant in the works of Hieronymus Bosch (1453–1516). In the centuries that followed, curiosity was supplemented by a desire for rationalization. The portrayal of the extra-ordinary body moved from the magical-religious dimension of myth to the scientific-descriptive one. The fascination with "diversity" and the attempt to rationalize it (thus bringing the monster back into order) are often combined in scientific works, such as *De monstris* by Fortunius Licetus

(1616) and *Monstrorum historia* by Ulisse Aldrovandi (1642). The baroque aesthetic mixes curiosity, a taste for the exotic, and merely aesthetic pleasure. It still featured monsters, by which is meant beings out of the norm, and therefore continued to arouse amazement.

Extra-ordinary bodies were included in architectural decorations, in gardens, even in the family portraits of the wealthiest families, where to host at least one person suffering from dwarfism was almost an obligation of high status. He or she would be incorporated in family portraits or even exhibited in a collection, as in the Agostino Caracci's *Triple Portrait of Arrigo, Pedro and Amon* (1599). The painting shows a man, Arrigo, at that time "owned" by the noble Farnese family, who has hypertrichosis. Arrigo is dressed as a "savage" and is shown in the company of exotic animals (a parrot, a monkey), a hunting dog, lapdogs, a "dwarf," and a "jester." The depiction of extra-ordinary bodies and nonhuman animals in the same scene conveys the idea of how these individuals were considered midway between animals and humans and therefore exhibited as curiosities and manifestations of luxury (Pedraza 2009).

A passion for the non-"normate" body also characterized the nineteenth century. Here, however, the magic/esoteric is relegated to certain figures (vampires, werewolves). A pseudo-scientific approach based on categorization and rationalization, linked to a growing need for classification and order, sparked interest in everything that, though human, is out of the norm. Monsters were now perceived as a "mistake of Nature," a breaking in the scientific rationality of modernity. For this reason, bodies whose forms seemed to transgress the social categories of gender (like the bearded woman) and human species (like the "mule-man" or the "crocodile man") become particularly sought after (Garland Thompson 2017).

The transition from amazing to ridiculous to pitiable is part of the contemporary representation of the extra-ordinary body. The posthuman approach aims at changing the perspective. Posthumanism internalizes the hybrid, the "cyborg," the monster, as its point of departure, to destabilize the limits and symbolic borders posed by the notion of the human as referred to in Western history, that is "white, male, heterosexual and propertied citizens, who would comply with institutionalized norms, as well as with ethnic, cultural, and physical characteristics" (Ferrando 2019, 4).

In children's literature, the dividing line between the human and the nonhuman animal is often called into question (Harrison 2018). In the *Harry Potter* saga (1997–2007), for instance, the moral relationship between human beings and other forms of life is a running motif. In particular, the author, J.K. Rowling, focuses on animal sentience, the relationship between human beings (or wizards) and the natural world, and also "monstrosity" (Dendle 2008). In the saga, boundaries between species are challenged by transformation, metamorphosis, hybridization. Body instability is a tenet, so much so that it makes "any stable sense of embodiment at best problematic and at worst impossible" (Harrison 2018, 327). At Hogwarts, instead of being fixed representations of human bodies, the portraits on the walls move and speak. Real bodies can appear and disappear, thanks to the cloak of invisibility. Witches and wizards, called Animagi, can take on the appearance of an animal at will. Harry Potter's body is transformed in a myriad of ways, so much that it may be considered posthuman itself (Batty 2015). To episodic and voluntary transformations linked to magic are added involuntary transformations like those imposed by Remus Lupin's lycanthropy. While Harry Potter's transformations are capable of accessing a nonhuman subjectivity, Lupin's werewolfism places him in the condition of not being recognized as human; indeed, in the *bestarium* studied by Harry as a textbook, it is written that "werewolves [...] have been shunted between the Beast and Being division for many years" (Batty 2015, 27). Like werewolves, the "merpeople" are also discriminated against because they cannot speak like human beings.

Becoming aware of this "speciesist" discrimination helps Harry to challenge not only "the viability and the desirability of being 'purely' human" (Harrison 2018, 326), but also the validity of a system that creates a hierarchy of bodies (and individuals) in accordance with a socially constructed norm.

9.2 Body order versus extra-ordinary bodies: *The Dolphin People* and *The Shape of Water*

The Shape of Water (2017) and *The Dolphin People* (2006) are two texts that, at first glance have very little in common. The first is a film directed by an award-winning filmmaker, Guillermo del Toro, which grossed $195,333,312 worldwide,[2] received 13 Academy Award nominations, and won four Oscars, including Best Picture. The second is a novel by a mysterious author who uses the name Torsten Krol, but whom no one has ever seen, not even his Australian publisher with whom he only communicates by mail. *The Shape of Water* is set in the US in the 1960s. It tells the love story of a mute young woman with an amphibious creature captured in the Amazon. *The Dolphin People* is set in the 1940s and tells the story of a German family traveling by plane who crash-land in the Amazon forest and get lost.

On closer examination, however, the two texts have much in common, starting with the Amazonian forest as "otherness" in spatial terms. In both cases, the plot establishes an opposition between, on the one hand, "order" and those who want to impose it, and on the other, those who respect and even admire the extra-ordinary body and its right to exist; therefore between anyone who wishes to defend order consider people with an "extra-ordinary" body as impure, eliminable, and expendable in scientific terms, and those who want instead to defend them. In both texts, the one who wants to impose order is the villain, and he is a character with all the features of the "human" in accordance with the criteria of Western humanism: he is a white, handsome, heterosexual male (Ferrando 2019) with a "normate" body. He wants to rectify the disorder or sacrifice it in the name of scientific progress. In this way the two texts question body binarism, and the male–female, human–animal, nature–culture dichotomies. What both texts fail to question is the geographical dualism between the West and the Rest (Hall 1992), since they associate order, or the pursuit of order, with the scientific (and militarized) reality of respectively, Germany and the US, and dis-order with the mysterious "elsewhere" of the Amazonian forest (a fact that is made more evident in *The Dolphin People* by the rather flat and stereotyped representation of the people living in the forest).

The Dolphin People is set immediately after the Second World War. The story is told by Erich, a boy of 16. His mother, Helga, is a war widow. She has decided to marry Klaus, the brother of her dead husband, and to move with him and her two sons (Erich and Zeppi, the younger brother) to Latin America. During the journey to their destination (the facilities of a company where Klaus found work as a physician), the plane crashes in the Amazonian forest. They are rescued by the Yayomi, a tribe of natives with no contact with the outside world. The Yayomi believe that the family are a family of dolphins that emerged from the waters of the river. Erich, Zeppi, Helga, and Klaus thus become "dolphin people." Gerhard, a German anthropologist who has spent 11 years with the natives and knows nothing about the war, helps the family, by acting as interpreter and cultural mediator.

The novel is presented as a coming-of-age story, with the result that Erich's first-person account becomes more articulate and detailed as the narrative unfolds. Described by Klaus as "a splendid specimen of Aryan boyhood," Erich voluntarily enters into a "going Indian" process (Baird 2018), gets an indigenous wife, thus losing his virginity, and gradually acquires local cultural traits (to the point of getting a tattoo on his face). Leaving Western culture behind, Erich gets rid of the nationalist and racist beliefs he was taught in Germany. Racial issues become apparent in his eyes. He allows memories of his Jewish friend to resurface, on whom, while in Germany, he had turned his back; he also remembers when his now-dead father confessed to him that he had not been a hero in the war and that he did not deserve the Iron Cross that Erich still so proudly flaunts round his neck. Equally clear in his eyes is the imbalance in gender relations, especially in the case of the Yayomi population.

As the plot unfolds, the other dolphin-people also become, or prove to be, very dissimilar from their initial image. Helga, the flawless mother, cannot accept the new situation she is in; she never washes, so as not to undress in front of others, starts to stink, goes crazy, and dies. Klaus, elegant and impeccable at the beginning of the novel, is revealed to be an SS officer on the run, proud to have exterminated hundreds of thousands of Jews; he is also a drug addict, who at first uses morphine, then survives a crisis of abstinence, and eventually starts abusing indigenous hallucinogens. Only little Zeppi remains faithful to his persona; described by Erich at the beginning of the novel as "the prettiest boy" he has ever seen, to the point of looking "like a girl," Zeppi indeed turns out to have breasts under the shirt.

In short, Zeppi has an extra-ordinary body that the group has to come to terms with. If modernity is order and cleanliness, a body that does not fall within the norm is a body that is not modern. Its place, therefore, is in the forest (the spatial "Otherness"), not in a civilized country. Hence, Klaus tells Zeppi: "You're better off here in the jungle with the Indians, with these … splendid appurtenances, than you would be in the city. The Yayomi may think nothing of them, nothing at all." Klaus, for his part, is modern and does not tolerate what he calls "mistakes of nature"; "abominations must be disposed of immediately," he says, while "nature's simple mistakes" can be corrected "with the scalpel." He cultivates the ambition to advance science through surgical experimentation (so much so that, when he was in the concentration camps, he detached the arm of a healthy Jew to transplant it onto another man). When he gets the chance, he tries to get Zeppi's body back in "order," and removes his male genitals, causing him to die.

Despite its symbolic importance as a space of otherness and disorder, in opposition to the order of modernity, the forest remains a setting. It is a barely described backdrop, a spatial container, not a place with which to relate. Hunting is practiced normally, animals are pets, and the forest is opposed to civilization in a painstaking repetition of a spatial binarism. However, for a few moments an unexpected simile brings the human world close to that of nonhuman animals' suffering. Regarding the scream of his dying mother, Erich says that it is "like the sound made by horses in the slaughterhouse back home as the hammer came smashing down between their eyes."

The Shape of Water is set in Baltimore in the 1960s. Elisa is a young woman who can hear but not speak on account of an injury she suffered as a child; she works as a cleaner in a laboratory. One day, a mysterious being is brought in; the creature is partially human (and male) but lives in water and has amphibious features. He comes from the Amazon, and there he is considered a god. A military man, Colonel Richard Strickland, brought the "monster" to the laboratory to vivisect him to benefit scientific knowledge. As in *The Dolphin People*, the villain in *The Shape of Water* is also a champion of "humanity" (male, white, heterosexual, with a "normate" body), and aims at the destruction of the extra-ordinary body.

Elisa teaches the monster how to speak with signs. As she does so, she falls in love with him and, with the help of her friends Giles and Zelda, succeeds in rescuing him and setting him free. Those who oppose the "normate" villain are all "marginalized" people: Elisa is mute, Zelda is a Black woman and Giles is gay. It is not difficult to understand with whom the director's sympathy lies, and the appeal of the contrast between the "human, social monster" hiding behind the "normate" body and all those whose identities are excluded, denied, or violated by him (Monteith 2018). Showing not only the possibility, but even the beauty of the love between Elisa and "the monster," "*The Shape of Water* embraces an alternative view of humanity that disengages the human from a particular set of physical traits" (Mejia 2018, 717), and perhaps goes beyond the very dimension of humanity itself. From this perspective, the monster does not need to turn into a human when he gets kissed by the princess, like the Frog Prince or the Beast in fairy tales: he already looks beautiful, at least in the eyes of Elisa and her friend Giles (even if he eats Giles' cat, reasserting the fact that he is also an animal). In this way, Del Toro's film dialogues with posthumanism, and questions the liberal humanist vision that considers the exceptionality of the human being in relation to nature as a granted asset (Mejia 2018).

9.3 Between human and animal: *Truismes*

Truismes (1996, Marie Darrieussecq) is a novel about the transformation of a body. At the time of its publication, the book was so "astonishingly successful" (Rodgers 2000) that it became a literary event (Depoux 2004). By June 1999, it had already sold 450,000 copies and had been translated into 34 languages (Rodgers 2000).

In French, the word "truismes" indicates something that is very obvious; but it is also possible to use the term about things that have to do with a "truie," the French word meaning sow. The wordplay in English is untranslatable. The English translation of the novel has a different title, *Pig Tales: A Novel of Lust and Transformation*, which is less intriguing but more explicit about the book's topic. The novel is the autobiography of a Parisian masseuse, very successful as a prostitute, whose body slowly turns into that of a sow. What is so evident as to be self-evident, as suggested by the wordplay of the French title, is the metaphor of the woman's flesh as "pork meat" (Cottille-Foley 2002). More specifically, the narrator is the victim of a patriarchal society that abuses women, but she has internalized society's values (so much so that she becomes a sow) and only evaluates herself in accordance with the way others look at her (Rodgers 2000).

However, the text goes far beyond this truism. It deals with the theme of writing, an inherently human act (even if the narrator apologizes at the beginning of the book for her *écriture de cochon*, literally "pig-writing"). Moreover, from a posthuman perspective, it questions the relationship between humans and nonhuman animals and investigates the different perception of the world that an animal like a pig has, compared to a human being. It also addresses the theme of the death of pigs in slaughterhouses from the opening epigraph (a passage by the Norwegian author Knut Hamsun). It describes a dystopian society politically dominated by a single leader. Eventually, in contrast with the urban reality's devastated environment, it offers the human-sow hybrid the "green space" of the "forest" (thus demonstrating the persistence of the forest as a symbolic space of "otherness," disorder and non-modernity).

The story's primary focus is the narrator's body, with its changes in sensations and perceptions and its hybrid situation (Damlé 2012). As written by Anat Pick (2011, 79) "From Ovid to Kafka, narratives of the transformations of species have served as a vehicle for discussing human identity, failings and flaws." If in *The Dolphin People* the transformation from human to animal occurs only in the imagination, in *Truismes* the metamorphosis is material. It is so intense and immersive as to make writing difficult for the narrator, as holding the pen causes terrible cramping to a limb that is no longer a human limb. The process is not sudden, as in Kafka, but slow and painful. First, there is hunger and an inevitable increase in physical roundness. The flesh becomes elastic and full. Then comes an aversion to pork and the disappearance of menstruation. Eventually, hair and additional nipples appear.

Highly appreciated for her beauty and the perfection of her body, the narrator becomes increasingly fat, vulgar, and animal even in her sexual practices (*une vraie chienne*, a real bitch). She begins to walk on all fours; her voice no longer articulates words. Her tastes change, rendering acorns and even dead bodies food of quality. Her sense of smell becomes much more powerful.

Throughout the process of transformation, the narrator maintains her human identity while also assuming an animal identity. Her body alternates between moments when it is still clearly human and moments when it is porcine. Her selves multiply, with no real opposition between her two identities (Rolls and Vuaille-Barcan 2009). This continuous transition breaks the classic dichotomy that exists between human beings and nonhuman animals, giving to the nonhuman animal an intense subjectivity. The metamorphosis is thus a metaphor used to underline the "violations that specifically 'animalize'" the narrator (Pick 2011, 83), but it is also a way to explore animality as a condition, introducing the idea of hybridity as a way to overcome the separation between human and nonhuman animals.

A dystopian setting in the background is controlled by an authoritarian regime dominated by a single leader, in which anyone not homogeneous in racial terms is liable to be expelled from the political community. The regime is violent, and, under its aegis, nefarious acts are carried out. Yet, with the same detachment and the same inversion of moral values with which she describes the violence imposed on her own body, the narrator offers a "normalizing" description of the dreadful acts that are carried out on others (Rodgers 2000).

In search of a final refuge, the story develops against a setting of different places of accommodation. In the beginning, there is the workplace in the perfumery where it is "normal" to prostitute oneself and where it is equally familiar for the shop manager, before giving you a job, to touch your breasts. Then there is the boyfriend's house; then, the search for a new shelter (a mental hospital, the city sewers, the crypt of a church, the house of another partner, who happens to be a werewolf). Throughout the narrator's search for a place to live, her animality increases, together with her desire to experience nature. In the countryside, she becomes intoxicated by the earth's scents, which enter her as if they were the entire planet: "cette odeur c'était comme si la planète entrait tout entière dans mon corps, ça faisait des saisons en moi, des envols d'oies sauvages, des perce-neiges, des fruits, du vent du sud" (that odor was like the whole planet entering my body, conjuring up in me seasons, flights of wild geese, snowdrops, fruits, south wind) (*Truismes*, 1996/*Pig Tales*, 1997).[3]

Her mother's house, the last place where the narrator seeks shelter, is a farm with an abattoir attached. So, the physical metamorphosis is accompanied by the animal terror of the slaughter. The topic is introduced in the opening epigraph, a quotation from *Benoni*, a novel by Knut Hamsun (1908), which returns in the narrative, about the killing of a hog.

The narrator manages to escape and takes refuge in the forest. She decides to end her journey and spend most of her time as a sow (and even meets a handsome and virile boar). If not with an actual voice (because the narrator can no longer speak after her transformation), the sow keeps a narrative voice using her *écriture de cochon*. In this way, "the text also brings forth how the ontology of nonhuman animals needs to be rethought" (Koponen 2017, 66). Again, literary fiction has given a subordinate the chance to speak, however, in this case too, the writer enjoys a very different existential condition (Marie Darrieussecq, who at the time of writing the novel was 27, had already completed her PhD, and taught literature at the University of Lille). There is no lack of clichés, such as the contrast, typical of dystopia, between the violent, urban, human world and the haven and peace offered by the forest, and therefore the nature–culture geographical dualism, which seems to be one of the most difficult to avoid. However, the text has the merit, among others, of introducing a reflection on how nonhuman animals perceive the world: life seen from a sow's viewpoint is different, yet her suffering is not.

Notes

1 www.treccani.it/vocabolario/ricerca/mostro/
2 www.boxofficemojo.com/title/tt5580390/?ref_=bo_se_r_1
3 The excerpt, from the novel *Truismes*, Marie Darrieussecq, is quoted with the permission of the publishing house P.O.L.

Bibliography

Baglioni, Igor (2013). Note alla terminologia e al concetto di "mostruoso" nell'antica Grecia. In Baglioni, Igor (ed.), Monstra. Costruzione e percezione delle Entità Ibride e Mostruose nel Mediterraneo Antico. *Thiasos. Rivista di archeologia e architettura antica* 2, 15–32. www.edizioniquasar.it/sku.php?id_libro=1934

Baird, Robert . (2018) Going Indian: Discovery, adoption, and renaming toward a "true American," from Deerslayer to Dances with Wolves. In Bird, Elizabeth S. (ed.), *Dressing in Feathers: The Construction of the Indian in American Popular Culture* (pp. 195–209). Routledge.

Batty, Holly (2015). Harry Potter and the (Post)human Animal Body. *Bookbird: A Journal of International Children's Literature* 53(19), 24–37. [https://doi.org/10.1353/chq.2018.0037]

Coppola, Giulio (2018). Arrivano i mostri; il mostruoso e l'ibrido tra antico e moderno. Un'esperienza didattica. *Mosaico* V, 65–79. www.liceofedericoquercia.edu.it › stories › pdf › mosaico

Cottille-Foley, Nora (2002). Métaphores, métamorphoses et retournements symboliques dans Truismes de Marie Darrieussecq: mais qui finit à l'abattoir?. *Women in French Studies* 10(1), 188–206. https://doi.org/10.1353/wfs.2002.0034

Damlé, Amaleena (2012). Posthuman Encounters: Technology, Embodiment and Gender in Recent Feminist Thought and in the Work of Marie Darrieussecq. *Comparative Critical Studies* 9(3): 303–318. https://doi.org/10.3366/ccs.2012.0065

Dendle, Peter (2008). Monsters, Creatures, and Pets at Hogwarts: Animal Stewardship in the World of Harry Potter. In Heilman, Elizabeth E. (ed.), *Critical Perspectives on Harry Potter* (pp. 175–188). Routledge.

Depoux, Anneliese (2004). La fabrique de l'événement littéraire: le cas de Truismes. *Communication & langages* 142(1), 71–83. www.persee.fr/doc/colan_0336-1500_2004_num_142_1_3306

Ferrando, Francesca (2019). *Philosophical Posthumanism*. Bloomsbury.

Garland Thompson, Rosemary (2017). Introduction: From Wonder to Error. A Genealogy of Freak Discourse in Modernity. In Garland Thompson, Rosemary (ed.), *Freakery: Cultural Spectacles of the Extraordinary Body* (pp. 1–19). New York University Press.

Hall, Stuart (1992). The West and the Rest: Discourse and Power. In Hall, Stuart, & Gieben, Bram, *Formations of Modernity* (pp. 275–332). Polity Press and Open University.

Haraway, Donna (1985). Manifesto for Cyborgs: Science, Technology, and Socialist Feminism in the 1980s. *Socialist Review* 80, 65–108 (also as *A Cyborg Manifesto: Science, Technology, and Socialist-Feminism in the Late Twentieth Century*. In Haraway, Donna (ed.), *Simians, Cyborgs, and Women*, Routledge, 1991, and in Haraway, Donna, *Manifestly Haraway*, University of Minnesota Press, 2016).

Harrison, Jen (2018). Posthuman Power: The Magic of Hybridity in the Harry Potter Series. *Children's Literature Association Quarterly* 43(3), 325–343. https://doi.org/10.1353/chq.2018.0037

Koponen, Päivi (2017). Animal Dystopia in Marie Darrieussecq's Novel Truismes. *Humanities* 6(65). https://doi.org/10.3390/h6030065

Li Causi, Pietro (2013). Mostri propriamente detti e creature paradoxa. Un tentativo di classificazione. In Baglioni, Igor (ed.), *Monstra. Costruzione e percezione delle Entità Ibride e Mostruose nel Mediterraneo Antico,* Roma. 53–67. *Thiasos. Rivista di archeologia e architettura antica* 2, 53–67. www.edizioniquasar.it/sku.php?id_libro=1934

Mejia, Silvia (2018). How Does Human Look? The Monster as the Ultimate Other in The Shape of Water. *Proceedings of the 5th World Humanities Forum* (pp. 717–723). Busan.

Monteith, Stephanie (2018). Queer as Fish. Love and Monstrous Bodies in 'The Shape of Water'. *Medium* 11. https://medium.com/@s.monteith1066/queer-as-fish-love-and-monstrous-bodies-in-guillermo-del-toros-the-shape-of-water-a60df1982928

Pedraza, Pilar (2009). *Venus barbuda y el eslabón perdido*. Siruela.

Pick, Anat (2011). *Creaturely Poetics: Animality and Vulnerability in Literature and Film*. Columbia University Press.

Rodgers, Catherine (2000) Aucune Évidence: Les Truismes De Marie Darrieussecq. *Romance Studies* 18(1), 69–81. https://doi.org/10.1179/ros.2000.18.1.69

Rolls, Alistair, & Vuaille-Barcan, Marie-Laure (2009). Une seule ou plusieurs femmes-truies? Une lecture virtualisante de Truismes de Marie Darrieussecq. *Australian Journal of French Studies* 46(1–2), 31–44. https://search.informit.com.au/documentSummary;dn=307146879834189;res=IELHSS≥

Sebenico, Sara (2005). *I mostri dell'Occidente medievale: fonti e diffusione di razze umane mostruose, ibridi ed animali fantastici*. EUT Edizioni Università di Trieste.

4

Reframing carnism

10 Carnism in popular culture

10.1 Introducing "carnism"

"Parents, teachers, official or friendly, doctors, not to speak of the powerful individual whom we call 'everybody', all work together to harden the character of the child with respect to this 'four-footed food', which, nevertheless, loves as we do, feels as we do, and, under our influence, progresses or retrogresses as we do" (On Vegetarianism, 1901). The quote comes from a short article by the French geographer Élisée Reclus (1901, 317). Reclus was an anarchist geographer (Ferretti 2011) known for the originality of his positions in social geography (Clark 1996). In his extensive scientific output, only this essay tackles the issue of vegetarianism. All the same, it should be enough to give him fame. Reclus tries to figure out why people eat animals and gets right to the core of the problem. Why do people kill animals to eat them, even if they know that animals feel and love as humans do? Because "everybody" does it.

Contemporary research has identified a series of psychological mechanisms that work in this way. Popular culture plays an essential role in normalizing the belief system, in which eating certain animals is considered ethical and appropriate (Joy 2010, 30). In these next chapters, we will investigate this role.

Before Reclus, many other intellectuals argued against the habit of eating meat. One of the most brilliant advocates of the vegetarian diet in the classical world was the Greek philosopher Plutarch, who lived between 40 to circa 120 CE. In his short essay *On the Eating of Flesh* (Περὶ σαρκοφαγίας – *De esu carnium*), published in his *Moralia*, the reasons given have to do with the beauty of animals, the desire to avoid waste, and the desire to avoid unnecessary suffering. Before Plutarch, other Greek authors, among whom Pythagoras and his followers, had practiced vegetarianism, however, Plutarch was the first not to link vegetarianism with the concept of the transmigration of souls (Spencer 1993). As remarked by Newmyer (1995), Plutarch not only transcended the religious motivations of his predecessors but also approached the topic in a "modern" way, taking into account the fact that animals are sentient creatures, and that human beings do not have the right to rob them of their lives with cruelty. He first declares absurd the belief of those who claim that the eating of meat has a natural origin. Human beings do not have beaks, claws, or teeth similar to those of predatory animals. He then reproaches the behavior of most humans, who do not kill the animals themselves but wait for others to do it; nor do they eat raw meat but cover it with spices of every kind and cook it to hide the taste of the blood. Finally, he says, eating meat is a habit embedded, like a hook, in our craving for pleasure. In short, Plutarch's arguments against eating meat have much in common with those in the current debate, i.e. we justify the fact that we eat meat because we consider it "natural," "normal," and "nice" (Piazza et al. 2015). Besides, eating animals is a difficult habit to eradicate.

The idea that eating meat is not natural was taken up again at the beginning of the nineteenth century by the English poet Percy Bysshe Shelley in a short essay with the title *Vindication of Natural Diet* (1813), where he affirms, among other things, that "Man resembles no carnivorous animal." After listing the merits of an animal-meat-free diet, he adds, moving on to a theme that today we call "sustainability": "The most fertile districts of the habitable globe are now actually cultivated by men for animals, at a delay and waste of aliment absolutely incapable of calculation" (Shelley 1813, 20). He concludes by saying that he and his wife (Mary Shelley) have been vegetarians for eight months when writing the booklet and are doing very well.

Accepting as normal that certain nonhuman animals may be killed for food while others are so dear to us that we make sacrifices for them, and even shed tears at their death, requires several steps. As Reclus writes, it is a mechanism triggered in childhood. Indeed, we are affected by the opinion of "everybody" from childhood: not only "parents, teachers, official or friendly, doctors," but more generally speaking "the powerful individual whom we call everybody" (Reclus 1901, 317) teaches us to see (some) animals as food.

This powerful "everybody" has that widespread and unquestioned mindset that Antonio Gramsci, in 1929, called "common sense." Common sense is a stratification of ideas, points of view, and ideologies, which becomes effectual, not because it is true, but because it has been embedded in social behavior throughout history. In Gramsci's words, "common sense" is a "conception of the world which is uncritically absorbed by the various social and cultural environments in which the moral individuality of the average man [sic] is developed" (1971, 419). There is not only one "common sense," of course. The taken-for-granted world changes over time. What yesterday was "commonsensical" may no longer be "commonsensical" today and vice versa. Moreover, within different cultures, a plurality of "common senses" may coexist.

To denaturalize common sense, Gramsci proposes that it should be historicized, that is to say relativized diachronically, along the historical path of human societies. However, he proposes that, from the cultural and social points of view, it should be relativized synchronically for the different stratifications (classes and social groups) of the same society. Gramsci's concept of common sense, together with the power–knowledge nexus, falls within the Foucaldian idea of "discourse," which is the basis of critical geopolitics. We can, therefore, address the "habit of eating meat because everybody does it" and because "it is natural" as a "discursive practice" and try to understand how popular culture conveys it.

The belief system in which eating certain animals is considered ethical and appropriate is called "carnism" by Melanie Joy (2010, 30). Like any other commonsensical attitude, carnism is a choice, even if it seems to be self-evident. Indeed "eating animals is a cultural choice and habit rather than a biological requirement, as plant-based diets have been shown to be nutritionally adequate and often healthier in terms of disease prevention and longevity" (Freeman 2014a, 193). As a "philosophical choice," carnism is based on an accepted belief system, not a biological necessity. Nevertheless, it appears to be a "non-choice," because "It's just the way things are" (Joy 2010); that is, it is made "invisible" by "common sense." No name exists to identify meat-eaters. They are "Everybody." Or almost everybody. Those who choose not to do so need be identified with a name. So, they are recognized as "vegetarians," or "vegans" (veg*ns). If "carnism" is a part of our common sense, to understand why "everybody" has come to think that eating certain animals (and not others) is right and appropriate, it is necessary to make an effort to "denaturalize" it both from a cultural and historical point of view. Melanie Joy (2010, 96) speaks about the "Three Ns of Justification" that act as mental and emotional blinders. The three Ns tell us that eating meat is "normal, natural, and necessary." Sometimes, it is also "nice" (so now the literature on the subject tends to speak of "the 4Ns of justification") (Piazza et al. 2015; Hopwood and Bleidorn 2019).

Indeed, eating meat is considered "normal," because it is a social norm; as such, it is not "merely descriptive — describing how the majority of people behave," it is "prescriptive, dictating how we ought to behave." Eating meat is also "natural," because, it is said, it is part of our eating habits. Humans have been eating meat for at least two million years. "But," Joy adds, "to be fair, we must acknowledge that infanticide, murder, rape, and cannibalism are at least as old as meat-eating, and are therefore arguably as natural." According to some researchers, and Plutarch's old philosophy, "humans are anatomically and physiologically adapted for a diet comprised primarily or entirely of plant foods" (Masson 2009, 30). However, meat is considered a "necessary" component of the human diet though contemporary dieticians have shown that this is not the case; proteins can also be taken from plant products, and one can live healthily on a plant-based diet.

In addition to the "Three Ns of Justification," Melanie Joy highlights how carnism rests on three cognitive mechanisms that lead to a distortion of perceived reality. She calls them the "Cognitive Trio" and terms them "objectification," "deindividualization," and "dichotomization." Objectification means that sentient animals are seen as things. It passes first and foremost through language: dead animals become meat, the calf becomes veal, and the pig pork. The word "veal" has no resonance in English. It comes from the translation of the French *veau* (calf); "would offend some people's sensibilities to be asked if they would like to be served calf for dinner. It's not dissimilar from 'pork,' which comes from the French *porc*, meaning pig" (Joy 2010, 145).

Objectification also comes through legislation, which makes animals human property. Deindividualization means that animals we eat are not unique individuals, just numbers. Lastly, dichotomization means dividing animals into categories such as inedible/edible to justify eating certain species (or the animals raised in a certain way) and spare others. Dichotomization (or "splitting," Masson 2009) calls for self-justification and leads to "compassionate carnism" (Joy and Tuider 2016). It may refer to the distinction made between animals bred in the right conditions, in good farms (which can be eaten), and animals bred in bad conditions (a thing not morally acceptable). Such distinctions, which are clearly "social constructs" (Leroy and Praet 2015, 68), are manifold and can refer, in a very fluid way, either to the species of the animals, or to their age, how they have been bred, and how they have been slaughtered (Arcari 2020); overall, they help the well-meaning consumer to forget that the ultimate end for all these animals is death. To this "cognitive trio," Jeffrey Moussaieff Masson (2009) adds another psychological mechanism, again echoing Plutarch, which is "denial." Denial helps us to choose not to consider in their entirety the information available to us about the death of animals and their suffering. As remarked by Noélie Vialles (1987, 125) "slaughtering tends to be a somewhat 'unpopular' subject: no one wants to know about it." Therefore, we do not kill the animals we eat but have others do it; we prefer slaughterhouses to be far away from our cities, and called by names such as meat plants, or packing plants; we cover meat with spices of every kind, as Plutarch said, to hide the taste of the blood.

All these psychological mechanisms mutually support each other, sustaining what constitutes the "common sense" about animal consumption. In consequence, like any form of common sense or ideological system, they deserve to be questioned. To take a critical approach, we can return to the concept of "discourse," in other words, we can identify carnism as the dominant discourse of contemporary ethics about food and proceed to its deconstruction. A long historical course leads to the construction of carnism as a discourse to which the roles of history, cultural traditions, the meat industry, the media, and popular culture are all contributing factors, as well as being profoundly interconnected and mutually supportive.

From a historical viewpoint, the passage from the ritual killing of a single animal to the standardized killing of the modern meat industry (Reed 2014) plays a vital role in the "deindividualization" and "abstraction." In a "sacrificial" offering, the animal is decorated for the purpose, and its death is celebrated as a sacred moment. In ancient Greece, the slaughter of animals for religious sacrifice was usually performed in temples, so "the death of an animal was a gift rather than a violation" (Young Lee 2005, 7). In contrast, in the modern meat industry, a bovine animal is slaughtered every twelve seconds (Pachirat 2011). As Reed (2014, 2) points out, the public slaughtering of animals, albeit in the form of religious sacrifice, has been progressively banned by societies as an emblem of violent ferocity; at the same time, the systematic abstraction of "meat" from "the animal" has been implemented. Thus, the more democratic the consumption of meat becomes, that is, the more frequent and numerous the killings are, the more the slaughtering and packaging of meat are removed from the sight and settlements of consumers, helping them to build their "denial." The progressive removal of the "animal origins of meat" is seen as "part of the process of civilization" (Buscemi 2014, 952). The process also involves cooking, where the preparation and cutting of the meat can hide the shape of the animal, and extends to butchering, meat preparation, and purchase.

The slaughterhouse emerged as a new institution in the early nineteenth century as part of a broader transition from an agricultural to an industrial economy, accompanied by increased urbanization, technological development, and an interest in public hygiene. An entirely modern invention, the centralized slaughterhouse was a political response to the public's growing lack of tolerance for "dirty" slaughter practices, corresponding to changing standards in social hygiene and the fear of meat-borne diseases. Within the slaughterhouse, in Europe and the Americas, animals' slaughter was rationalized in line with capitalist imperatives. In parallel with the introduction of hygiene regulations, the slaughterhouse's expulsion from the city streets served to save consumers from confronting the animals' animality. A Royal Decree issued in Italy in 1890 stated that slaughterhouses should be located "away from the city center and educational institutions" to avoid the annoyance caused by the "screams of the animals." The spatial obliteration of death was added to the animals' linguistic and normative objectification and the denial of their individuality. This "geographical" euphemism is still implemented, not only through the removal but also through the anonymization (White 2015) of the sites. "The reality is mostly obscured by [...] the creation of a thin veneer of civility surrounding human-animal relations, embodied largely by language tricks, isolation of death camps, and food preparation routines that artfully disguise the true origins of flesh-food" (Wolch and Emel 1998, XI).

Together with objectification and deindividualization, dichotomization too has a long cultural history. In contemporary Western societies, characterized by an unprecedented level of meat consumption, and by a very intense process of industrialization in the manufacturing of animal products, there is an extraordinary contrast in the treatment given to meat animals and to pets, which have never been so numerous or pampered. This dichotomy is nothing new, and it refers to more than the distinction between pets and farm animals. For cultural and religious reasons, different food traditions variously differentiate between the animals that can be eaten and those that cannot. Broadly, a classification could be made between animals that are bred to be eaten, work animals, pets, animals considered traditionally impure, and "sacred" animals. However, how these species are classified is relative culturally (some people eat dogs, others consider it criminal to eat cows) and historically. Cats were considered edible in different regions of Italy until the last century; rabbits and horses are still eaten, but fewer people consider it acceptable. According to the anthropologist Marvin Harris, the author of *Good to Eat: Riddles of Food and Culture* (1998), different eating habits are simply a consequence of the need to adapt one's diet to environmental diversity.

10.2 The "meat paradox" and beyond: how the hegemonic dietary discourse of carnism is produced and reproduced by advertising

Hiding slaughterhouses, placing them far from the places where we live, or making them anonymous, hiding them "in plain sight" (White 2015), is not enough. As Ralph Waldo Emerson wrote (1860, 5), "however scrupulously the slaughterhouse is concealed in the graceful distance of miles, there is complicity." The awareness of this "complicity," i.e., the psychological conflict between the dietary preference for meat and the awareness of the suffering inflicted on farm animals, has been defined as "the meat paradox" (Loughnan, Haslam and Bastian 2010; Bastian and Loughnan 2017). Harming others is incoherent with the vision of oneself as a moral person. Meat consumption makes meat-eaters confront a view of themselves that is unfavorable: How can I be a good person while eating meat? Meat-eating, they say, "conflicts with deeply held moral principles, yet people seek to justify these self-serving behaviors so as to protect their own interests" (Bastian and Loughnan 2017, 1). Indeed, we know that "Animal husbandry is one of the greatest sources of suffering in the world…," but, at the same time, "Hypocrisy feels less bad, less threatening, when in a group. If all of us are doing something bad, it can't really be that bad, right?" (Shaw 2019). Julia Shaw calls this attitude "the psychology of group-think" (Shaw 2019).

In the case of meat-eating, there are two stakeholders: the consumers, who enjoy eating meat, and the meat producers, who gain from selling it. In the middle, there are the animals. If consumers are at risk of suffering from the "cognitive dissonance" (Festinger 1957) that arises from "the attitudinal inconsistency between both desiring meat and having concerns over animal welfare" (Bastian and Loughnan 2017, 2), meat producers must intervene. They have to induce consumers to overcome their moral reticence and disconnect the image of meat from the death of the animal. In this way, the animal becomes an "absent referent" (Adams 2004). So, "in addition to our own attempts to justify meat-eating, corporations double down to make it easier for us to do so" (Shaw 2019).

The meat industry may resort to different strategies and communication euphemisms to mitigate this conflict. While making the slaughterhouse anonymous, hidden in plain sight (White 2015), is one of them, another consists in making all the process of "meat production" appear "scientific," and "mechanically clean." The slaughterhouse becomes a rationally organized activity in the killing of animals. The system is perfected to be increasingly mechanical, rational, and effective. It even changes its name; it is no longer a slaughterhouse but becomes a "beef factory," or a "meat processing factory," or, in North America, a "packaging factory" (Philo and MacLachan 2018, 101). To reduce empathy, "killing" and "slaughtering" are now called "harvesting" (Kunst and Hohle 2016). The slaughterhouse thus becomes a "perfectly neutral" institution in the face of serial death, since "the machine has no emotion" (Young Lee 2005, 25).

Advertising can employ other communication strategies. The meat industry's publicity may represent meat as the "right" thing to eat or even the "necessary" thing to eat. In the 1940s, a famous Leo Burnett campaign for the American Meat Institute presented the consumer with the picture of a piece of raw red meat against a red background, accompanied by a knife and fork (Buscemi 2018), and slogans like "This is life," or "Sure, you're right in liking meat." In this way, the campaign not only suggested the importance of eating meat to human life (meat is "life," it is a human "necessity") but also the fact that it is "natural" and also "right." It is also "normal," because the image of the raw meat helped the consumer overcome the conflict between the desire for meat and its potentially repulsive image.

Advertising campaigns must suggest to consumers that meat is not just a natural, normal, and "right" food. It is "nice" and "everybody" likes. In many Italian television advertisements of packed meat products, the slogans repeat that "everybody" wants just that product, to reassure viewers of its excellent quality, and at the same time to reaffirm that eating this particular product is the "normal" choice. In a 1960s commercial, the leading character is a cowboy who, when it comes to the final shootout, shows that he has canned meat in his bandoleer instead of pistols. On seeing the meat, "everybody goes to eat with him" rather than duel. Similarly, in a contemporary advertisement, a child is seen snacking on ham and gives some to his toy dinosaur, mirroring the slogan "everyone wants a slice."

"How we present, prepare, and talk about meat increases willingness to eat meat by reducing empathy and disgust" (Kunst and Hohle 2016, 758). So, the meat industry displays meat in a sanitized way, plastic-wrapped and already chopped into pieces, "making it hard to think that it came from an animal at all" (Shaw 2019). Meat can also be presented already cooked and turned into a food that in no way recalls its animal origin.

Sometimes, animal images are used in advertising materials. In this case, they are often shown as cartoon figures, using *Kindchenschema* ("baby schema") (Lorenz 1971), which is the set of juvenile features that help humans feel affection for animals: large eyes, bulging craniums, and round features. This kind of representation, supposed to give the impression that meat comes from imaginary animals (Shaw 2019), is a sort of visual euphemism — "cutification" (Grauerholz 2007). Not only do animals appear "cute enough to eat," but they are smiling in a friendly way, they look happy, they even appear "willing accomplices in their own slaughter and consumption" (Grauerholz 2007, 349). This communication strategy is quite common in Europe, where nowadays it is used mostly by small producers, butchers, delicatessens, and dairy producers. In this regard, it is worth remembering the extraordinary popularity of *Mucca Carolina* (Carolina Cow), a cartoon with an exaggeratedly cutified cow as the main character, which was used in the 1960s to promote a big dairy company. For dairy products, in a greenwashing effort that is as brazen (given the conditions of intensive animal farming) as it is widespread, animals are not only cutified; they are usually drawn against a background of green pastures.

Using "green pastures" to associate the animal with a pleasant environment means using a visual propaganda technique called "transfer." Generally speaking, we can compare advertising to political propaganda: in both cases, communicators have a message and want to convince their listeners and viewers of its validity. In the meat industry, many advertisement messages are often similar to those conveyed by the most common propaganda devices, identified in 1939 (Lee & Lee 1939) by the Institute for Propaganda Analysis (IPA) (Sproule 2001). Eat meat because "everyone does it" ("bandwagon"); do it because "it's good, it's right, it's healthy" ("glittering generalities"); do it because he (or she) does it too ("testimonials"); do it because it is healthier, it has more protein, it has fewer calories (than what?) ("cardstacking").

10.3 "Let's have a hot dog": meat eaters (and veg*ns) in popular culture

The carnist discourse, and more generally, the cultural attitude that allows people to exploit animals and their products (milk and eggs, as well as meat) for their pleasure, is not only the result of a cultural tradition or the communication strategy of corporations. In this regard, Melanie Joy (2010) highlights the role of the news media. The horrific conditions of animals subjected to CAFO (Concentrated animal feeding operations) and the daily slaughter of millions of farm animals do not make the news. Therefore, there is an omission of information. Animals are commodified in news media; their emotional perspectives are not represented; they are deindividualized (Freeman 2009).

Moreover, meat consumption, which appears to be central to many global customs and celebrations, is regularly presented as "normal"; on the occasion of special festivities (e.g., Thanksgiving Day in the US, Easter in Italy), it even becomes "a social obligation" (with the turkey or the lamb on the table). So, not only do the mainstream "media fail to challenge the system," they also maintain its invisibility and "reinforce the justification for eating meat" (Joy 2010, 103). At the same time, veganism, which falls outside of carnist discourse, is generally presented as a choice contrary to common sense. As demonstrated by a research conducted by Cole and Morgan (2011), the press often represents vegans as ascetic, faddish, oversensitive, and hostile characters.

Popular culture plays a similar role. TV and movie characters eat, and so make dietary choices, and even novels are full of references to food. The art of painting does not disdain filling its works with dead animals, whether they are game or slaughtered cattle (see the pictorial tradition of still lifes with dead game, or the work of Renato Guttuso, *La vucciria*, 1974, which has a quarter of an ox in the foreground). Even popular music happens to exalt food of animal origins, and in the US, there is a country band called *Red Meat*.

Given the prevalence of omnivorous people, portraying people who eat animal products seems justified. "Meals containing animal derivatives are ubiquitous in American mainstream media. Almost every family dinner scene, every celebration party depicted, every romantic date represented, every situation that has to do with food, usually involves meat, dairy, or/and eggs" (Grande 2017). In contemporary movies, it is common for the characters to be seen eating eggs, to having a hot-dog together, or going to a fast-food restaurant for a hamburger. In American movies, if there is spaghetti on the table, it is spaghetti with meatballs; if there is pizza, it is pepperoni pizza. Sometimes, eating a certain kind of meat is also a sign of national belonging or identity. Spaghetti with meatballs is a cliché for Italo-American culture and Italo-American restaurants in the US, together with a red check tablecloth and a waiter with a moustache. Eating hamburgers is a synonym for being American, not only in the US but worldwide (Ohnuki-Tierney 1997). In *Harold & Kumar Go To White Castle* (2004, Danny Leiner), two boys of Asian origin (Harold and Kumar) cross New Jersey to reach their favorite burger place and dispel the discomfort caused by racism. Eating meat can also be synonymous with strength (as Walt Whitman asks himself, "How is it I extract strength from the beef I eat?" in the poem *Song of Myself*, 1855), gender, and social status. For instance, *Babette's Feast* is a short tale by Karen Blixen (1950) that was later adapted into a film (1987, Gabriel Axel), in which a Parisian cook is forced to leave France incognito and finds refuge in a small village in Norway (in Denmark, in the movie) . When, out of gratitude, she organizes a banquet for her host family, the menu is made sumptuous by the presence of certain animal products (turtle broth and quails). Conversely, the choice, or obligation, to eat only vegetables is often used as a metaphor to denote a situation of penitence or punishment.

Altogether, literature, cinema, and TV replicate the dominant way of viewing animals and food "through an anthropocentric and carnistic lens," so they "reflect this carnonormativity" (Freeman 2014a, 196). Generally speaking, vegetarian and vegan (i.e., veg*an) characters are quite rare. If there are any, they are hybrid monsters (as in *Frankenstein*, in Mary Shelley's novel, 1818) or genetically engineered post-human creatures, like the Crakers in Margaret Atwood's *Oryx and Crake* (Quinn 2018). If they are boys, they are often a "soyboy" (Gambert and Linné 2018).

A notable exception is represented by Libereso, the main character in the story "Adam, One Afternoon" by Italo Calvino (in Italian, "Adamo, un pomeriggio," in the collection *Ultimo venne il corvo*) (1949). Libereso is a 15-year-old boy, a gardener in a villa, who wants to make a present to a 14-year-old kitchen maid and proposes as gifts the little animals living in the garden, which are all his friends. Libereso says to the girl: "'We don't eat the flesh of dead birds or animals', 'Why, are you always in Lent?', 'What do you mean?', 'Well, what do you eat then?', 'Oh, all sorts of things, artichokes, lettuces, tomatoes. My father doesn't like us to eat the flesh of dead animals'" (*Adam, One Afternoon*, 1949 [1958]).

Then he adds that his father, in the evening, reads Elisée Reclus' books aloud to his family (in the English translation, 1958, quite amazingly, Elisée Reclus' name is replaced by Kropotkin's) (Guest 2017). Libereso is the son of an anarchist and his name means "freedom." Maria-nunziata, the girl, is instead a victim of all sorts of rules. She is disgusted by all that Libereso offers her as a gift, but at the same time she is attracted to it. In this case, vegetarianism does not question masculinity, but it does certify how much the character is out of the chorus (and outside the geopolitical discourse of his time). By the way, Libereso is not a fictional character. Libereso Guglielmi was a young man who worked in the Calvino family garden and was a pupil of Calvino's father, a famous botanist. Libereso's father, in turn, was an anarchist, very active on the Italian scene (Ferrua 1977).

The trope "real men eat meat" is taken to extremes in the satirical animated sitcom *South Park*, in the episode *Fun with Veal* (2002, season 6, episode 4). One of the characters (Stan) decides to become vegetarian after discovering that veal is made of baby cows, but it is so labeled to make it palatable for consumption; visiting a veal farm, he also discovers that the calves are chained to prevent them from moving, to ensure the tenderness of their muscle tissue. To prevent calves from being killed, he manages to trigger an ecoterrorist action. However, one of the other kids (Cartman) tells him, "if you don't eat meat, you become a pussy." And, indeed, Stan develops a strange form of illness and begins to grow vaginas on his face. With this ambivalent attitude, *South Park* on one hand advocates animal rights and censures meat industries for manipulating markets and consumers (Zekavat 2019), and on the other risks conveying the message that "interest in animals makes one a 'giant pussy'" (Fathallah 2015, 1311).

If they are not monsters, mutants, or anarchists, vegetarian characters are girls. An association between vegetarianism and femininity was already highlighted at the end of the nineteenth century with two feminist utopian novels: *Mizora* (1880, Mary E. Bradley) and *New Amazonia: A Foretaste of the Future* (1889, Elizabeth Burgoyne Corbett). *Mizora* describes a women-only utopian society in which people live on fruit and artificial meat. In *New Amazonia*, set in a similarly utopian society populated only by women, everyone is vegetarian. Nowadays, the connection is still present. In contemporary films and TV shows, vegetarian girls are usually brilliant, like Ellie Wood in the movie *Legally Blonde* (2001, Robert Luketic), Darlene Conner in the TV sitcom *Roseanne* (1988–98), Phoebe in *Friends* (1994–2004), and Lisa in *The Simpsons* (1989–). Sometimes, they are also potentially laughable figures (Jallinoja, Vinnari, and Niva 2019). Phoebe, for instance, is portrayed as a bit eccentric, confronted with a world where "good people" eat meat, which legitimizes carnism by emphasizing its normalcy. However, Lisa is not only vegetarian but also an activist who tries to convince her community to adopt similar positions. Unfortunately, people greet her pleas with laughter. However, this reinforces her message: even if her crusade is unsuccessful, she, who advocates reasonable arguments, is right while others, who do not understand them, are retrograde. It might be said that, in contrast with *Friends*, "*The Simpsons'* producers are using laughter as a way to ridicule morally inconsistent social norms rather than just encourage continued carnistic conformity" (Freeman 2014, 6).

Generally speaking, popular culture tends to represent veg★n characters as "killjoys" that ruin the pleasure of others by highlighting the unethical nature of their behavior (Grant and MacKenzie-Dale 2016). An exchange of lines from the Spanish television series *White Lines* (2020) provides a good summary of this stance, albeit ironic since the speaker is a drug dealer. At first, he defines "healthy, vegan, crudités" as "the scourge of the 21st century." And then he adds "no more meat because f…ing cows contaminate the Earth. No more rock and roll, man. No more Barbie girls, no more fun" (*White Lines*, season 1, episode 7).

Bibliography

Adams, Carol J. (2004). *The Pornography of Meat.* A&C Black.

Arcari, Paula (2020). *Making Sense of 'Food' Animals.* Basingstoke.

Bastian, Brock, & Loughnan, Steve. (2017). Resolving the Meat-Paradox: A Motivational Account of Morally Troublesome Behavior and its Maintenance. *Personality and Social Psychology Review* 21(3), 278–299. https://doi.org/10.1177/1088868316647562

Buscemi, Francesco (2014). From Killing Cows to Culturing Meat. *British Food Journal* 116(6), 952–964 https://doi.org/10.1108/BFJ-11-2012-0288

Buscemi, Francesco (2018). *From Body Fuel to Universal Poison: Cultural History of Meat: 1900-The Present.* Springer.

Clark, John (1996). The Dialectical Social Geography of Élisée Reclus. In Light, Andrew, & Smith, Jonathan M. (eds.), *Philosophy and Geography I. Space, Place, and Environmental Ethics* (pp. 117–142). Rowan & Littlefield.

Cole, Matthew, & Morgan, Karen (2011). Vegaphobia: Derogatory Discourses of Veganism and the Reproduction of Speciesism in UK National Newspapers 1. *The British Journal of Sociology* 62(19), 134–153. https://doi.org/10.1111/j.1468-4446.2010.01348.x

Emerson, Ralph Waldo (1860). *The Conduct of Life*, H.O. Houghton.

Fathallah, Judith (2015). Won't Somebody Please Think of the Children? Or, South Park Fanfic and the Political Realm. *Journal of Youth Studies* 18(10), 1309–1325 https://doi.org/10.1080/13676261.2015.1039972

Ferretti, Federico (2011). The Correspondence between Élisée Reclus and Pëtr Kropotkin as a Source for the History of Geography. *Journal of Historical Geography* 37(2), 216–222. https://doi.org/10.1016/j.jhg.2010.10.001

Ferrua, Pietro (1977). Il sostrato sanremese nella narrativa di Italo Calvino. *Italica* 54(3), 367–380. [https://doi.org/10.2307/478157]

Festinger, Leon (1957). *A Theory of Cognitive Dissonance.* Stanford University Press.

Fonseca, Rui Pedro (2015). The Presence of Carnism on Portuguese Television. *Open Journal of Social Sciences* 3, 48–55. [https://doi.org/10.4236/jss.2015.38004]

Freeman, Carrie P. (2009). This Little Piggy Went to Press: The American News Media's Construction of Animals in Agriculture. *The Communication Review* 12(1), 78–103. https://doi.org/10.1080/10714420902717764

Freeman, Carrie P. (2014a). Lisa and Phoebe, Lone Vegetarian Icons: At Odds with Television's Carnonormativity. In Macey, Deborah A., Ryan, Kathleen M., & Springer, Noah J. (eds.), *How Television Shapes Our Worldview: Media Representations of Social Trends and Change* (pp. 193–212). Lexington Books.

Freeman, Carrie P. (2014b). *Framing Farming: Communication Strategies for Animal Rights.* Brill.

Gambert, Iselin, & Linné, Tobias (2018). From Rice Eaters to Soy Boys: Race, Gender, and Tropes of 'Plant Food Masculinity'. *Animal Studies Journal* 7(2), 129–179. https://ro.uow.edu.au/asj/vol7/iss2/8

Gramsci, Antonio (1971). *Selections from the Prison Notebooks* (translated and edited by Quintin Hoare and Geoffrey Nowell Smith). International Publishers.

Grande, Luis Alejandro (2017). *Speciesism and Carnism in Media Animal Consumption in TV and Film.* University Of Miami. Unpublished Master Thesis. www.researchgate.net/profile/Luis_Grande5/publication/334151046_Speciesism_and_Carnism_in_Media_Animal_Consumption_in_TV_and_Film/links/5d1b702e458515c11c0c4673/Speciesism-and-Carnism-in-Media-Animal-Consumption-in-TV-and-Film.pdf.

Grant, Juawana, & MacKenzie-Dale, Brittni (2016). Lisa Simpson and Darlene Conner: Television's Favorite Killjoys. In Castricano, Jodey, & Simonsen, Rasmus R. (eds.), *Critical Perspectives on Veganism* (pp. 307–329). Palgrave Macmillan.

Grauerholz, Liz (2007). Cute Enough to Eat: The Transformation of Animals into Meat for Human Consumption in Commercialized Images. *Humanity & Society* 31(4), 334–354. https://doi.org/10.1177/016059760703100404

Guest, Bertrand (2017). Environmental Awareness and Geography: Reading Reclus Ecocritically? In Specq, François (ed.), *Environmental Awareness and the Design of Literature* (pp. 69–89). Brill Rodopi.

Harris, Marvin (1998). *Good to Eat: Riddles of Food and Culture.* Waveland Press.

Hopwood, Christopher J., & Bleidorn, Wiebke (2019). Psychological Profiles of People Who Justify Eating Meat as Natural, Necessary, Normal, or Nice. *Food Quality and Preference* 75, 10–14. https://doi.org/10.1016/j.foodqual.2019.02.004

Jallinoja, Piia, Vinnari, Markus, & Niva, Mari (2019). Veganism and Plant-Based Eating: Analysis of Interplay between Discursive Strategies and Lifestyle Political Consumerism. In Boström, Magnus, Micheletti, Michele & Oosterveer, Peter (eds.), *The Oxford Handbook of Political Consumerism* (pp. 157–179). Oxford University Press.

Joy, Melanie (2010). *Why We Love Dogs, Eat Pigs, and Wear Cows: An Introduction to Carnism.* Red Wheel/Weiser/Conari.

Joy, Melanie, & Tuider, Jens (2016). Foreword. In Castricano, Jodey, & Simonsen, Rasmus R. (eds.), *Critical Perspectives on Veganism* (V–XV). Springer.

Kunst, Jonas R., & Hohle, Sigrid M. (2016). Meat Eaters by Dissociation: How We Present, Prepare and Talk About Meat Increases Willingness to Eat Meat by Reducing Empathy and Disgust. *Appetite* 105, 758–774. https://doi.org/10.1016/j.appet.2016.07.009

Lee, Alfred, & Lee, Elizabeth Briant (1939). *The Fine Art of Propaganda.* Harcourt, Brace and Company.

Leroy, Frédéric, & Praet, Istvan (2015). Meat Traditions. The Co-Evolution of Humans and Meat. *Appetite* 90, 200–211. https://doi.org/10.1016/j.appet.2015.03.014

Lorenz, Konrad (1971). *Studies in Animal and Human Behavior.* Harvard University Press.

Loughnan, Steve, Haslam, Nick, & Bastian, Brock (2010). The Role of Meat Consumption in the Denial of Moral Status and Mind to Meat Animals. *Appetite* 55(1), 156–159. https://doi.org/1016/j.appet.2010.05.043

Masson, Jeffrey Moussaieff (2009). *The Face on Your Plate: The Truth about Food*. W.W. Norton & Company.

Newmyer, Stephen T. (1995). Plutarch on the Moral Grounds for Vegetarianism. *The Classical Outlook* 72(2), 41–43. www.jstor.org/stable/i40162291

Ohnuki-Tierney, Emiko (1997). McDonald's in Japan: Changing Manners and Etiquette. In Watson, James (ed.), *Golden Arches East: McDonald's in East Asia*, 2nd ed. (pp. 161–182). Stanford University Press.

Pachirat, Timothy (2011). *Every Twelve Seconds: Industrialized Slaughter and the Politics of Sight*. Yale University Press.

Philo, Chris, & MacLachlan, Ian (2018). The Strange Case of the Missing Slaughterhouse Geographies. In Wilcox, Sharon, & Rutherford, Stephanie (eds.), *Historical Animal Geographies* (pp. 100–120). Routledge.

Piazza, Jared, Ruby, Matthew B., Loughnan, Steve, Luong, Mischel, Kulik, Juliana, Watkins, Hanne M., & Seigerman, Mirra (2015). Rationalizing Meat Consumption. The 4Ns. *Appetite*, 91, 114–128. http://dx.doi.org/10.1016/j.appet.2015.04.011

Pick, Anat (2011). *Creaturely Poetics: Animality and Vulnerability in Literature and Film*. Columbia University Press.

Quinn, Emelia (2018). Monstrous Vegan Narratives: Margaret Atwood's Hideous Progeny. In Quinn, Emelia, & Westwood, Benjamin (eds.), *Thinking Veganism in Literature and Culture* (pp. 149–173). Palgrave Macmillan.

Reclus, Élisée (1901). On Vegetarianism. *The Humane Review* 1(4), 316–324.

Reed, Annette Yoshiko (2014). From Sacrifice to the Slaughterhouse: Ancient and Modern Approaches to Meat, Animals, and Civilization. *Method & Theory in the Study of Religion* 26(2), 111–158. www.jstor.org/stable/43907142

Shaw, Julia (2019). *Evil: The Science Behind Humanity's Dark Side*. Abrams.

Shelley, Percy Bysshe (1813). *A Vindication of Natural Diet*. Smith & Davy.

Spencer, Colin (1993). *The Heretic's Feast: A History of Vegetarianism*. Fourth Estate.

Sproule, J. Michael (2001). Authorship and Origins of the Seven Propaganda Devices: A Research Note. *Rhetoric & Public Affairs* 4(1), 135–143. www.jstor.org/stable/41939653

Vialles, Noélie (1987). *Le sang et la chair – Les abattoirs des pays de l'Adour*, Éditeur Maison des Sciences de l'Homme (1994, *Animal to Edible*. Cambridge University Press).

Young Lee, Paula (2005). The Slaughterhouse and the City. *Food and History* 3(2), 7–25. https://doi.org/10.1484/J.FOOD.2.301751

White, Richard J. (2015) Following in the Footsteps of Élisée Reclus: Disturbing Places of Inter-species Violence that are Hidden in Plain Sight. In Nocella, Anthony J. II., White, Richard J., & Cudworth, Erika (eds.), *Anarchism and Animal Liberation: Essays on Complementary Elements of Total Liberation* (pp. 212–230). McFarland Press

Wolch, Jennifer R., & Emel, Jody (eds.) (1998). *Animal Geographies: Place, Politics, and Identity in the Nature-Culture Borderlands*. Verso.

Zekavat, Massih (2019). Satire, Humor and Ecological Thought. *Neohelicon* 46(1), 369–386. https://doi.org/10.1007/s11059-018-00471-0

11 Engendering meat

11.1 Meat, myths, masculinity

The "meat-fun-Barbie girls" connection made in the TV series White Lines is maybe ironic, but it is not accidental. "The high value of meat is largely contingent upon its symbolic importance as a tangible representation of human control of, and superiority over, nature" (Fiddes 1991, 17). Metaphorically, meat is strength. So, as a symbol of power, it is not surprising that meat consumption is associated with other discursive constructions of authority, inside a schema where the "ideal subject" is "a human, male, meat-eating individual, actively 'possessing' nature and accepting the sacrifice inherent to eating flesh" (Derrida 1991, 114). To define this schema, which illustrates the "ideological basis for the predominance of meat-based diets in Western culture" (Flail 2011, 84), Derrida coined the word "Carnophallogocentrism" (1991).

This idea of "carnivorous virility" adds to the concepts of carnism and speciesism, a gender differentiation element, and focuses on the role of carnivorism in the making of "male subjectivity." Carol J. Adams addresses the point in *The Sexual Politics of Meat: A Feminist-Vegetarian Critical Theory* (1991) and *The Pornography of Meat* (2004). Adams connects meat-eating and the patriarchate, analyzing myths of the past (such as the Greek myth of Zeus and Metis) and contemporary literary works, especially feminist literature and vegetarianism. Erika Cudworth (2005) coins the term "Anthroparchy" to reference the system of power that connects anthropocentrism, patriarchate, animal exploitation, and the capitalist domination of nature.

The connection between masculinity and meat has a long history in different cultures. Again, there are ancient myths connecting strength, virility, and meat. Freeman (2014b) quotes the Egyptian myth of the sacred bull Apis, which symbolized fertility, courage, power, and even kingship. Then, there are male-dominated hunting activities. Since the Upper Paleolithic, in many different situations, men have devoted themselves to hunting and women to gathering. Some researchers explain this differentiation by referring to the dangers of hunting (Leroy and Praet 2015), but some deviations from the pattern did occur. They add that "focus should therefore not be on the sexual task division in pregiven gender terms, but rather on the emergence of gender within a specific socioecological context" (Leroy and Praet 2015, 10). Hunting was already subject to ritualistic control in prehistoric times, and different cultures developed different rules (Luke 2007). In many cultures, hunting soon became a pastime for the privileged and powerful. Even in the early Middle Ages, the aristocracy of the remnants of the Western Empire was often more interested in hunting than in agriculture, so much so as to impose areas to be used exclusively for hunting, where the production of food was forbidden.

Hunting played a key role in aristocratic education: it was exalted as a preparation or alternative to war, which kept boys away from more effeminate pleasures. Moreover, it allowed one to demonstrate the ability to understand the signs of nature and knowledge of natural history. The combination of rules and courage made it a highly refined pastime. Hunting practices were regulated on values of correctness and loyalty: examples are that a wounded animal had to be finished off, at the cost of following it inside a tangle of vegetation; it was forbidden to shoot at random into the herd; and that some animals could only be shot if they were moving (such as ducks, which otherwise were "sitting ducks," i.e., too easy prey), and others only if they were standing. Various elements accentuated the sexual symbolism of the hunt; usually, with the exception of the big felines, the coveted prey was the male of the species. Even the collecting of skins and especially of horns can be read and interpreted in this sense: the trophy room was, in fact, part of the aristocrats' homosocial space.

Hunting had a specific meaning in the colonial context. The importance of hunting, a prerogative usually reserved to the nobles, took on a central role in imperial ideology, so much so that in colonial Africa, it represented a focal point in the interaction between Africans and Europeans (Mackenzie 1988). Many high-ranking hunters disapproved of hunting as a livelihood connected with commercial activities or with Africans' survival techniques. The access of local inhabitants to hunting was progressively reduced, and areas reserved for the activity. For Europeans, hunting became a symbol of their position of domination and a class determinant. Only those who could afford it did so: the rich, the nobility, high-ranking officers. On the whole, hunting, when not associated with the need to provide food, demonstrated courage (and therefore virility) and skill.

The "erotic of men's predation" (Luke 2007) can still be found in the language used in contemporary hunting literature, where there is an abundance of expressions connecting the hunter to the prey with terms such as "romance" and "love." Sometimes they are even tinged with phallocentric sexuality. The hunting weapon becomes an extension of the hunter's body, which penetrates the animal's body, while the animal becomes an almost erotic object of desire. The mental association between hunting and sexual practice is highlighted in American everyday life by the slogan, recurring on caps and car stickers, "I Hunt Whitetail All Year Long." The image of a deer, framed by the viewfinder, on the left, and of a woman with her back squared in the same way, on the right, accompany the slogan (Luke 2007).

Animal killing can also be interpreted as "a central part of the rites of passage framing the transition to manhood" (Leroy and Praet 2015, 23). Killing animals without falling into childish sentimentalism can turn boys into men. The killing of animals as a sign of transition to manhood was still present in US popular culture of the twentieth century. The movie *The Yearling* (1946, Clarence Brown, based on a novel of the same title by Marjorie Kinnan Rawlings) is an excellent example. It tells the story of Jody, a lonely boy who lives in the Florida countryside with his family and adopts a fawn as a pet. The deer, once grown, destroys the family's fields and crops. So, Jody's mother decides to shoot it. She only wounds it (she is a woman, of course, and does not know how to shoot well!), and Jody himself has to finish off his beloved deer with a rifle shot. Only then does the mother, previously distant and seemingly devoid of affection, begin to show Jody the love he deserves.

11.2 *Real men don't eat quiche*: gender stereotypes and dietary habits in the media

Real men don't eat quiche is the title of a satirical book by Bruce Feirstein, published in 1982, where the author, pretending to teach his readers how to behave as "real men," makes fun of gender stereotypes. First of all, to be a real macho, a man should not be a "quiche eater." The cliché of the real man who eats meat is very powerful; it has historical roots linked to hunting traditions, and is continuously nourished by popular culture.

In TV shows, movies, and cartoons, meat consumption is often associated with masculinity; thus, very masculine characters are defined by the fact that they eat hamburgers and steaks. Meat is strength and being a red-meat eater increases muscle strength (as recurrently suggested by men's magazines such as *Men's Health*) (Stibbe 2004). In relation to this, Leo Burnett's campaign for the American Meat Institute has to be quoted again, since Burnett affirmed that "we convinced ourselves that the image of meat should be a virile one, best expressed in red meat [...] Red against red was a trick... but it was a natural thing to do. It just intensified the red concept and the virility and everything else we were trying to express" (Ewen 1998).

Moreover, meat is connected with sexual desire. On this, Carol J. Adams (1991, 210) quotes Elaine Showalter (1985, 129), who claims that in the late nineteenth century meat was "believed to be the fuel of anger and lust," while "a carnivorous diet was associated with sexual precocity, especially with an abundant menstrual flow, and even with nymphomania." So, women were not supposed to like meat, because it did not suit them (Belasco 2008). Indeed, many girls, fearing their sexuality, were often disgusted with meat, or even phobic about it (and, along with Eros, they also rejected male power) (Adams 1991, 214). Alternatively, they ate meat with sensual greed, especially if it was usually denied them. In this case, however, they were not represented as girls from good families. Elsie, in Arnold Bennett's *Riceyman Steps* (1923), is a young maid who performs a forbidden act: eating her employer's leftover steak. In

Elsie's eyes, as Mary Addyman (2017) points out, the steak is not just food, but a temptation of biblical significance, a seductive lure pushing her first to lick her fingers, then to lose control of her actions.

Thus, the gender-meat-sex nexus turns out to be very complex. Meat is related to virility, strength, and pleasure. It is also associated with the female body. There is an abundance of metaphors that link the female image to a piece of meat (so much that a dating site may be referred to as a "meat market"). Popular culture provides explicit examples. In Margaret Atwood's first novel, *The Edible Woman* (1969), Marianne, the lead character, stops eating when she begins to see the food on her plate come to life. This also happens with meat, which causes her a sudden sense of rejection and disgust. She manages to free herself from this form of mental illness when she also gets rid of her boring boyfriend Peter, by whom she feels somehow "eaten." Even more explicit than *Truismes* (1996, Marie Darrieussecq) in making a connection between sex and meat is *The Butcher* (1988, orig. *Le Boucher*), an erotic novella by the French writer Alina Reyes. It is the story of a young woman who works at a butcher's shop in the summertime and feels erotic when watching the butcher cutting up the meat. The narrator's gaze lingers on the butcher's actions, as if she wanted to identify herself with that piece of meat, penetrated skillfully by the man's blade. Following John Phillips' interpretation, the author "celebrates the erotic body, whether male or female, as fragmented and objectified like pieces of butcher's meat" (Phillips 1999, 196). The hyperbolic metaphor, however, can also open up a critical approach, which highlights the connection between gender constructions as well as processes of objectification of women as "meat."

The cliché of the male super-carnivore may emphasize his "barbarian" nature or achieve a comic effect by unexpectedly reversing the stereotype. In the French comic series *Astérix le Gaulois* (1959, René Goscinny and Albert Uderzo), at the top of the culinary preferences of the two main characters, the Gallic heroes Asterix and Obelix, are roast boar and hog. In the historical drama television series *Vikings* (2013–19, History Channel), pigs' heads and other parts of killed animals are often seen, and all the characters stuff down a lot of roasted meat. Moving on to the contemporary era, in one episode (2005, *The Competition*, season 4, episode 23) of the US TV sitcom *According to Jim* (2001–09, ABC), the main character and his friend Andy engage in the "annual hot-dog eating competition." They feel sure to win as they are men, but surprisingly a woman takes the prize.

11.3 The "cow" and the "boy" along the trails of the west(ern)

Mainstream popular culture not only takes for granted that people eat meat, but also glorifies animal-based lifestyles. From the bucolic poems of the Greek poet Theocritus in the third century BCE to the contemporary western genre, pastoralism and its practices (breeding, shepherding, transhumance) have been praised as important traditions and ways of life. Specifically, western cinema and literature celebrate cattle-breeding as one of the founding elements of US national identity and its workers (the cowboys) as national heroes. "Eat Beef – The West Wasn't Won on Salad" is a catchphrase on bumper stickers, mugs, and T-shirts.[1] In a few words, this statement succeeds in exemplifying "carnivorous virility," and "celebrating the genocide of Native Americans" (Adams and Calarco 2016, 41). It also helps us to understand how human-animal binarism (which lies at the basis of speciesism), and man-woman binarism (which lies at the basis of gender discrimination), are also related to space binarism, which opposes the civilized space of the colonizers to the rest of the world open to conquest.

Since the nineteenth century, the development of settlement colonies, in particular in the Americas, has guaranteed an abundance of meat on the tables of rich Europeans. The whole history of the American West is associated with meat production. Cattle, which had been brought to the American continent by the Spanish colonizers together with horses, soon roamed wild again. Thanks to the favorable context, they began to multiply.

The progressive occupation by farmers and settlers of lands inhabited by Native Americans, on the one hand, pushed the frontier progressively towards the ocean; on the other hand, it allowed the cattle industry to grow ever larger. "The American cattle culture was gestated in Texas" (Rifkin 1992, 68). Cowboys tamed the "Wild West" and turned it into a vast pasture for cattle. It was in this way that red meat became associated with this category of men who embody manhood in the United States (Rifkin 1992). The production and marketing of beef, and in general the need to organize the industry to supply the large markets in the East of the country, required transport systems, commercial hubs, and urban networks. This development led to the expulsion of Native Americans from their lands and the near extinction of bison, which were slaughtered by the million to reduce the food resources available to the native populations, to feed the railway construction workers, and to empty the grasslands.

This gloomy historical picture, marked by the genocide of a people (the Native Americans) and the almost total extinction of an animal species (the bison), would seem destined not to enter a people's collective memory, but to be relegated to oblivion. On the contrary, the conquest of the West has become the cornerstone of the nation's narrative (Slotkin 1998), with the men who made the care of livestock their profession its protagonists and heroes. In the epic elaboration of this myth, popular culture has played a primary role in a variety of forms, such as songs and ballads, oral stories, paintings, prints, dime novels, cinema, and traveling shows, all of which had the West as a background and the cowboy (or bison hunter) as the central character. In this way, the United States created a national epic for itself: the history of the frontier, a genre that forged its national image — the western — and an undisputed model of national masculinity — the cowboy. Some of these forms of popular culture, such as itinerant performances of the *Wild West Show*, progressively tailed off in popularity, while others increased to become productions for the masses. The cultural products that contributed more than any other to making the West the mythical landscape of the American national epic were the movies, so much so that between the 1920s and 1970s westerns represented about a quarter of Hollywood's entire output (Buscombe 1988).

The young herdsman of the Great Plains was often a poorly paid and poorly fed boy, who was sometimes unemployed during the winter months. Thus, the reality of the time was that few young men aspired to being a cowboy: it was certainly preferable to be a rancher. However, in the popular imagination the cowboy prevails over other figures who lived on the frontier (hunters, soldiers, railway builders). The nomadic herdsman became an established figure in the Great Plains during the last quarter of the nineteenth century. It was at that time that the process of territorial conquest was closing and urban and industrialized America was beginning to miss it, which led to the events inspired by the conquest of the West entering American popular culture. The cowboy thus emerged as the figure best able to personify the nostalgia for a mythical past and uncontaminated "nature." As from the 1880s, Buffalo Bill's traveling *Wild West Show* put the figure of the cowboy on stage, helping to cement him as a mythical character. Next came the paintings of Frederic Remington and Charles M. Russell, and then, above all, Owen Wister's novel *The Virginian: A Horseman of the Plains* (1902), which, by promoting the cowboy to the rank of absolute protagonist of the West, elevated him to the role of the last American hero. With Wister's descriptions, the cowboy acquired the physical traits that he would later maintain over time: he was tall, pleasant-looking, and possibly blue-eyed (although in reality, he was often Afro-American or Latin).

In some movies, the figure of the cowboy merges with that of the gunfighter. In a typical western structure, exemplified by such films as George Stevens' *Shane* (1953) and Clint Eastwood's *Pale Rider* (1985), and then seen in hundreds of variations, the cowboy comes to town, helps honest but fearful citizens fight the baddies, then, after showing the community how to get by on its own, goes back to the emptiness where he came from (Wright 1977). In many other films, however, the cattle are the main focus, when an equally iconic plot describes the romance and skills involved in moving them from one place to another (often from the plains to the market). The typical cattle-drive story departs from Texas with as destination such places as Dodge City and Abilene, and states like Kansas or Montana, and is often accompanied by the coming-of-age tale of one or more youngsters. In cattle-drive movies such as *Red River* (1948, Howard Hawks and Arthur Rosson), *Cattle Drive* (1951, Kurt Neumann) or *The Cowboys* (1972, Mark Rydell), the perilous journey into adulthood of the characters overlaps with the journey of the cattle to their final destination (the meat market). Thus, while for the former, the "natural" reaching point is represented by adulthood, for the latter, the equally natural reaching point seems to be, as is taken for granted, the slaughterhouse.

Worth mentioning here is the television series *Rawhide* (1959–65, CBS), which focuses entirely on the "getting to the market" of the cattle and its dangers. The series is still remembered today for its theme music, the song *Rawhide* (1958, Ned Washington and Dimitri Tiomkin), which was brought back to fame by the cult movie *The Blues Brothers* (1980, John Landis). The word rawhide means untanned leather, or whip. The lyrics say that the calves must be made to move, even if they disapprove, not try to understand them, just make them move. Indeed, they must get to the market, i.e., end up in the slaughterhouse. For the cowboy instead, at the end of the ride, his true love will be waiting.

Such lyrics are perhaps not surprising, when considering that the song was written in 1958. The song's lasting success is proof of the power of carnism, so the fact that a young animal has to be urged with the whip to go to its death not only does not horrify us (because the obvious end of the young animal is that it will be turned into steaks), it is even turned into a song that children hum. The song was featured with its original lyrics in *The Blues Brothers*. Later, it was revived in different cartoon movies. In *An American Tail: Fievel Goes West* (1991, Phil Nibbelink and Simon Wells), the *Blues Brothers*' version is performed by some animals, while Fievel (a mouse) travels in tumbleweed. In *Shrek 2* (2004, Dreamworks), the character Donkey sings it briefly, but with different words. In *Happy Feet 2* (2011, George Miller), the song returns with its original lyrics, albeit sung with an Australian accent by marching sea elephants.

11.4 Mastering carnonormativity: television cooking shows and reality formats

Percy Bysshe Shelley wrote this more than two centuries ago in one of his essays on vegetarianism: "It is only by softening and disguising dead flesh by culinary preparation that it is rendered susceptible of mastication or digestion, and that the sight of its bloody juices and raw horror does not excite intolerable loathing and disgust" (*A Vindication of Natural Diet*, 1813). A similar affirmation had been made, about eighteen centuries earlier, by Plutarch. In *On the Eating of Flesh* (Περὶ σαρκοφαγίας – *De esu carnium*), the Greek philosopher writes that "we are so refined in our blood-letting that we term flesh a supplementary food; and then we need 'supplements' for the flesh itself, mixing oil, wine, honey, fish paste, vinegar, with Syrian and Arabian spices, as though we were embalming a corpse for burial."[2]

As Noélie Vialles claims (1987), there are "zoophages" that love to recognize animals in what they consume. They are a minority. Most meat consumers belong to a category which, for etymological homogeneity, she calls "sarcophages." They consume only an abstract substance, where nothing is identifiable from the animal. The role of cooking, sometimes elevated to cuisine, therefore becomes fundamental.

In this respect, TV cooking shows, be they talent shows or cooking demonstrations, are another area of popular culture that exalts carnism. In Francesco Buscemi's words (2016, 331), culinary talent shows end up having a "carnivorous mission": they promote a form of consumption that is today regarded as harmful to our health, just as cigarette smoking was years ago. They hide the origin of animal products in sophisticated recipes and normalize their use, never questioning the practice of cooking animal products. Eating meat and other products of animal origin is the "normal" form of sustenance. Cooking, moreover, can be a great professional skill and even a true artistic expression. The use of animal products is celebrated and their consumption spectacularized for the benefit of the audience. This kind of show tends to associate masculinity with cooking (and with meat), since the people in charge are mostly male chefs in a professional backdrop (Packham 2016). The format of *MasterChef* (Shine Group) offers the best example with its combination of critically acclaimed chefs as judges, and non-professional chefs as competitors. The usually male chefs-judges are presented in a charismatic manner and typically act authoritatively towards the contestants. The show thus helps create the figure of the chef as a media star but also promotes a gender discourse in which masculinity implies power and hierarchical superiority, thus mixing, once again, carnism with gender issues.

Even though people seem to have less time to cook, food has become one of the staples of television fodder (Versteegen 2010, 448). Television food is served in all sauces. It seems to fit every schedule and format: cooking classes, contests in which chefs compete against each other, inter-restaurant competitions, shows in which incompetent restaurateurs strive to improve, and travel programs about specific food regions. This trend started in the United Kingdom, where the most successful culinary talent television format, *MasterChef*, was launched in 1990. From there it spread around the world. Different versions of *MasterChef*, adapted to local realities (Rimoldi 2015), are produced in more than 40 countries. Different kinds of show satisfy different audiences. The most traditional cooking program, which teaches viewers how to cook, is usually broadcast a little before lunchtime. A presenter, usually a woman, demonstrates the cooking typical of a housewife in a pseudo-domestic set. Another format, a competition between restaurateurs or chefs, such as Italy's *Quattro Ristoranti* (based on the German format *Mein Lokal, Dein Lokal*) and *MasterChef*, attracts a more varied audience and is often shown in prime time. *MasterChef Australia* (2009–) is one of the country's most popular programs and is exported as a "canned program" (with subtitles or dubbing) to about 30 other countries.

Cooking programs and culinary talent shows, which "normalize" the consumption of "traditional" meats, may suggest the consumption of less usual meats at home. With reference again to *MasterChef Australia*, sellers and suppliers said that, after the show's first few seasons, the consumption of unfamiliar foods such as quails and pheasants had increased throughout the country.[3]

Sometimes, competitors are asked to deal with "unusual" food preparations, as occurred during one episode of the Portuguese *MasterChef* when contestants were asked to prepare a lamb that had not yet been cut up. To the competitors "the animals were presented [...] with their whole body (with head, eyes, teeth)" (Fonseca 2015, 49); in this way, the "absent referent" (that is, the dead animal beyond the food) (Adams 1991) was brought back, somehow implying that, for a cook, accepting the death of the animal is inevitable and "necessary." Even causing the death of the animal can be shown as necessary and inevitable: in an episode of the English edition (2010), competitors were asked to cook crabs by throwing them alive into boiling water. The strength of the carnist message was reinforced by the attitude of a competitor, a girl of Indian origin, who at first claimed to be vegetarian and never to have killed an animal in her life, but who then stated that she was ready to do it, somehow demonstrating the need to adapt to what everyone does (Rimoldi 2015).

In 2019, the desire to adapt to the growing sensitivity towards food of animal origin led to the invitation of a vegan chef in an episode of the UK edition. The competition between the contenders was therefore focused, for once, on the challenge of producing a plant-based menu. In an episode of the Italian edition in 2017, the two teams were confronted with the preparation of a vegan menu of raw food. These exceptions are so rare that they confirm the rule: for now, carnonormativity reigns on television networks.

However, new tendencies are emerging, which somehow seem to be a reaction to the growing attention to animal rights. On the one hand, the search for "ethical" meat seeks to make consumers aware of the conditions experienced by farmed animals. On the other, there is a "new carnivorism," the reaction of those who show contempt for ethical objections to the use of animals for food and who boast their right to eat meat, offering "new visibility to slaughter" (Parry 2010). For instance, the BBC's production *Kill It, Cook It, Eat It* (2007–11) was a television series that followed "the journey of animals from the pasture to the plate." The show echoed the idea that those who eat meat are generally detached from the reality of meat production and voluntarily ignore the death necessary to put it on their plate (Parry 2010), and aimed to recast the concept of turning "the repugnant into [the] acceptable" (Pick 2016, 95) by pointing out the importance of animal welfare and the need to choose "good meat," i.e., meat produced from animals that are not factory-farmed. Similarly, the Australian SBS documentary *For the Love of Meat* (2016) aims to induce the public not to give up meat but to consider purchasing animals farmed fairly. Both programs present the dichotomy between animals that are good to eat and those that are not (Joy 2010; Arcari 2020) while normalizing the use of animals as food (Arcari 2017). Conversely, the British reality show *Meat the Family* (Channel 4, 2020–) straightforwardly questions the dichotomy between animals for food and animals as pets. In the show, meat-consuming families are given farm animals (broilers, piglets, lambs) to raise for three weeks as pets. At the end of that time, they have to decide whether to send them to a sanctuary (but in that case, they have to become vegetarians) or let them go to the slaughterhouse. In that case, they will have to eat them. In the first episode, both families sent their adopted chickens and piglets to the slaughterhouse, shocking the audience.[4]

The Butcher is an American competition reality series (History Channel, 2019–) that is very explicit in its "new carnivorism" (Armstrong and Potts 2014, 179). In the show, competitors are required to show off their ability to dissect animals, which sometimes are very unusual (a nutria, an alligator, a bison), and often using even more unusual tools (such a tomahawk). In this way, the show focuses on the lack of connection between urban meat consumers and animals, contrasting traditional and "authentic" hunting practices with contemporary methods of industrial meat production (and again suggesting the possibility of "splitting" between animals that are good to eat and animals that should not be eaten).

Notes

1 The saying was launched by a 1990 North Dakota Beef Commission promotional campaign ("Beef...because the West wasn't won on salad"). https://www.barrypopik.com/index.php/new_york_city/entry/eat_beef_the_west_wasnt_won_on_salad
2 https://penelope.uchicago.edu/Thayer/E/Roman/Texts/Plutarch/Moralia/De_esu_carnium*/1.html
3 www.smh.com.au/entertainment/masterchef-fans-drive-specialty-boom-20100726-10smk.html
4 www.theguardian.com/tv-and-radio/2020/jan/09/meat-the-family-tv-doesnt-get-creepier

Bibliography

Adams, Carol J. (1991). *The Sexual Politics of Meat: A Feminist-Vegetarian Critical Theory*. Continuum.

Adams, Carol J. (2004). *The Pornography of Meat*. A&C Black.

Adams, Carol J., & Calarco, Matthew (2016). Derrida and The Sexual Politics of Meat. In Potts, Annie (ed.), *Meat Culture* (pp. 31–53). Brill.

Addyman, Mary (2017). Conclusion "All Else Is Vain, but Eating Is Real": Gustatory Bodies. In Addyman, Mary, Wood, Laura, & Yiannitsaros, Christopher (eds.), *Food, Drink, and the Written Word in Britain, 1820–1945* (pp. 207–220). Routledge.

Arcari, Paula (2017). Perverse Visibilities? Foregrounding Non-Human Animals in "Ethical" and "Sustainable" Meat Consumption. *The Brock Review* 13(1), 24–53. https://doi.org/10.26522/br.v13i1.1317

Arcari, Paula (2020). *Making Sense of "Food" Animals*. Basingstoke.

Armstrong, Philip, & Potts, Annie (2014). The Emptiness of the Wild. In Garry, Marvin, & McHugh, Susan (eds.), *Routledge Handbook of Human-Animal Studies* (pp. 168–181). Routledge.

Belasco, Warren (2008). *Food: The Key Concepts*. Berg.

Buscemi, Francesco (2016). The Carnivorous Mission of the Celebrity Chef. In Castricano, Jodey, & Simonsen, Rasmus R. (eds.), *Critical Perspectives on Veganism*. (pp. 331–348). Palgrave Macmillan.

Buscombe, Edward (1988). The Western: A Short History. In Buscombe, Edward (ed.), *The BFI Companion to the Western* (pp.15–54). Andre Deutsch Ltd.

Cudworth, Erika (2005). *Developing Ecofeminist Theory: The Complexity of Difference*. Palgrave Macmillan.

Derrida, Jacques (1991). Eating Well, or the Calculation of the Subject: An Interview with Jacques Derrida. In Cadava, Eduardo, Connor, Peter, & Nancy, Jean-Luc (eds.), *Who Comes After The Subject?* (pp. 96–118). Routledge.

Ewen, Stuart (1998). Leo Burnett: Sultan Of Sell. *Time* 152(23), 92–95.

Feirstein, Bruce (1982). *Real Men Don't Eat Quiche*. New English Library.

Fiddes, Nick (1991). *Meat: A Natural Symbol*. Routledge.

Flail, Gregory James. (2011). Why "Flexitarian" was a Word of the Year: Carnophallogocentrism and the Lexicon of Vegetable-Based Diets. *International Journal of Humanities and Social Science* 1(12), 83–92.

Fonseca, Rui Pedro (2015). The Presence of Carnism on Portuguese Television. *Open Journal of Social Sciences* 3, 48–55. https://doi.org/10.4236/jss.2015.38004.

Freeman, Carrie P. (2014a). Lisa and Phoebe, Lone Vegetarian Icons: At Odds with Television's Carnonormativity. In Macey, Deborah A., Ryan, Kathleen M., & Springer, Noah J. (eds.), *How Television Shapes Our Worldview: Media Representations of Social Trends and Change* (pp. 193–212). Lexington Books.

Freeman, Carrie P. (2014b). *Framing Farming: Communication Strategies for Animal Rights*. Brill.

Joy, Melanie (2010). *Why We Love Dogs, Eat Pigs, and Wear Cows: An Introduction to Carnism*. Conari Press.

Joy, Melanie, & Tuider, Jens (2016). Foreword. In Castricano, Jodey, & Simonsen, Rasmus R. (eds.), *Critical Perspectives on Veganism* (v–xv). Springer.

Leroy, Frédéric, & Praet, Istvan (2015). Meat Traditions. The Co-Evolution of Humans and Meat. *Appetite* 90, 200–211. https://doi.org/10.1016/j.appet.2015.03.014

Luke, Brian (2007). *Brutal: Manhood and the Exploitation of Animals*. University of Illinois Press.

Mackenzie, John M. (1988). Chivalry, Social Darwinism and Ritualised Killing: the Hunting Ethos in Central Africa up to 1914. In Anderson, David, & Grove, Richard H. (eds.), *Conservation in Africa. Peoples, Policies and Practice* (pp. 41–62). Cambridge University Press.

Packham, Charley (2016). A Woman's Place is in the Kitchen? The Relationship between Gender, Food and Television. In Bradley, Peri (ed.), *Food, Media and Contemporary Culture* (pp. 83–102). Palgrave Macmillan.

Parry, Jovian (2010). *The New Visibility of Slaughter in Popular Gastronomy*. Unpublished MA thesis. University of Canterbury.

Phillips, John (1999). *Forbidden Fictions Pornography and Censorship in Twentieth-Century French Literature*. Pluto Press.

Pick, Anat (2016). Animal Rights Films, Organized Violence, and the Politics of Sight. In Tzioumakis, Yannis, & Molloy, Claire (eds.), *The Routledge Companion to Cinema and Politics*. (pp. 113–124). Routledge.

Rifkin, Jeremy (1992). *Beyond Beef: The Rise and Fall of the Cattle Culture*. Dutton.

Rimoldi, Luca (2015). How to Show a National Cuisine: Food and National Identities in the Masterchef Kitchen. *Academic Journal of Interdisciplinary Studies* 4(2), 257–257. https://doi.org/10.5901/ajis.2015.v4n2p257

Shelley, Percy Bysshe (1813). *A Vindication of Natural Diet*. Smith & Davy.

Showalter, Elaine (1985). *The Female Malady: Women, Madness, and English Culture, 1830–1980*. Pantheon Books.

Slotkin, Richard (1998). *Gunfighter Nation: The Myth of the Frontier in Twentieth-Century America*. University of Oklahoma Press.

Stibbe, Arran. (2004). Health and the Social Construction of Masculinity in Men's Health Magazine. *Men and Masculinities* 7(1), 31–51. https://doi.org/10.1177/1097184X03257441

Versteegen, Heinrich (2010). Armchair Epicures: The Proliferation of Food Programmes on British TV. In Gymnich, Marion, & Lennartz, Norbert (eds.), *The Pleasures and*

Horrors of Eating: The Cultural History of Eating in Anglophone Literature. (pp. 447–464). V&R Unipress.

Vialles, Noélie (1987). *Le sang et la chair – Les abattoirs des pays de l'Adour*, Éditeur Maison des Sciences de l'Homme (1994. *Animal to Edible*. Cambridge University Press).

Wright, Will (1977). *Sixguns and Society: A Structural Study of the Western*. University of California Press.

12 Carnonormativity and its discontents

12.1 Cracking carnonormativity

According to Comolli and Narboni (1969, 11), "every film is political, inasmuch as it is determined by the ideology which produces it (or within which it is produced, which stems from the same thing). The cinema is all the more thoroughly and completely determined because unlike other arts or ideological systems its very manufacture mobilizes powerful economic forces in a way that the production of literature (which becomes the commodity 'books') does not — though once we reach the level of distribution, publicity and sale, the two are in rather the same position."

The authors are telling us that all texts (films, but also books) are related to an ideological position; the vast majority follow it, some oppose it, either because of the content or the form, or both. Specifically, the majority are so "imbued [...] with the dominant ideology in pure and unadulterated form" that they become "the unconscious instruments of the ideology which produces them" (Comolli and Narboni 1969, 13). Some texts instead attack the ideology. They may be politically committed in terms of the message, or formally resistant in their expressive form. Lastly, some texts seem to fit in fully with the dominant ideology but have "cracks" in them that may contradict the system.

Comolli and Narboni's analytical frame was later revised and summarized by Stam, Burgoyne, and Flitterman-Lewis (2005). Among other categories, they distinguish "dominant films," thoroughly imbued with dominant ideology; "resistant films," which attack dominant ideologies on the levels of both the signified and signifier; "formally resistant films," which do not have an openly political message, but practice formal subversion; and "fissure films." Fissure films superficially belong to the dominant cinema but provide a "crack" by means of a detail, a single scene, or a dialogue. Even if they seem to be integrated into the system and the ideology, such movies can help to dismantle the system from within. There is also "live cinema," which may critically depict contemporary events and sometimes also question the traditional mode of representation.

Comolli and Narboni's frame about the possible relations between a text and the dominant ideology was first formulated within a Marxist framework. However, by broadening the perspective from capitalism to patriarchy, and to speciesism, a similar frame helps analyze other sides of the hegemonic discourse that we may call "anthroparchy" (Cudworth 2005). In the following paragraph, I will try to analyze popular culture texts regarding carnism as ideology and to problematize representations of nonhuman animals and consumption practices.

As seen in the previous chapter, mainstream popular culture is generally so imbued with carnism as the dominant dietary discourse that it becomes its "unconscious instrument." Overall, this "carnonormativity" (Freeman 2014a) contributes to normalizing the consumption of animal products for the audience.

"Resistant" fictional texts, which can be interpreted as specific "politically committed" attempts to go against the carnist imperative, are few. A "resistant" film in content and form is *Il potere* (*Power*) by the Italian director Augusto Tretti (1971). As the title suggests, the film is an attack on power, which is examined in a historical arc ranging from prehistory to modernity, in all its forms (first of all religious, then political, military, and economic). Although the film does not advocate animal rights, animals repeatedly appear as victims of power. In the initial scene, a hen, chased by a multitude of men and women, ends up captured and plucked; in the conquest of the West, there is talk of the extermination of natives and bison; in the modern world, 100,000 egg-laying hens are shown crammed together on an industrial farm. The director shows hens "forced to wear glasses to avoid cannibalism, while the owner praises to his guests the qualities of the Super-Egg, actually a nothing full of water and without yolk, emblem of food in our society, beautiful to see, but without taste and nutritional values" (Lanzarotti 2010). In addition to the content, the film also appears "resistant" from a formal point of view. It is shot in black and white, with non-professional actors playing all the roles; the sets are declaredly theatrical; the three representatives of power (who illustrate the logic in the different historical phases) wear animal masks. Even the soundtrack, which is cacophonic and falsely triumphant, contributes to creating a sense of despondency in the spectator. In an interview, the director declared that he wanted to "ridicule the formal structures of consumer cinema, its dazzling colors, beautiful photography, luxury, and opulence" (Zambetti and Giacci 1973).

Less challenging in the form, but resistant in the content is *La Grande Bouffe/The Big Feast* (1973, Marco Ferreri). It is a French-Italian movie about four friends who eat so much that they die (*bouffe* is the French word for binge). This grotesque film is a satire of capitalist and bourgeois consumerism, not a direct attack on carnism. However, the "binge" involves the consumption of a large quantity of meat. The meat is delivered in its animal form. During the delivery scene, one of the characters dances with a calf's head; a second counts what is unloaded and comments in a sardonic way: "five innocent lambs!"; the third states, "you'll see how exciting it is to dissect the meat," the fourth declares the party open (a curious beginning, marked by the presence of so many dead animals). The scene returns the "absent referent" (the innocent lamb, the calf) to view, while the association between sexuality and meat consumption is somehow satirically reaffirmed, suggesting a generally critical message about meat consumption and anthroparchy.

Carnage: Swallowing the Past (2017, Simon Amstell) offers an explicit and direct attack on the carnist discourse, although, again, this is a film that is only resistant in content. *Carnage* is a film of fiction produced by the BBC in 2017. Set in the year 2067, it is constructed as if it were a documentary (it can be considered a "mock documentary" or mockumentary) and depicts a utopian future where everyone is a vegan. The younger generations do not understand how human beings were able to feed on animals in the past. The elders, who ate meat and cheese in the past, are full of guilt. They find themselves in therapy sessions, like an AA meeting, where they confess their mistakes. The film intelligently contemplates the historical steps that have led to this transformation of the world. In the intermediate "Age of Confusion" it was realized that even pets, such as cats and dogs, are carnivorous and unsustainable. It then demonstrates "how quickly such normalized social ideas as eating meat can be rejected" (Mills 2017, 179).

In the case of "resistant" texts such as these it would be useful to deepen the research in order to understand the reactions of the audience. *La Grande Bouffe*, which violates all rules of decency, was a scandal at the time; it was booed at the 26th Cannes Film Festival and then, maybe because of this, was quite a success at the box office. Despite the booing, the film won the "International Critics' Award" at Cannes, where it was presented again, in a restored version, in 2013. In the meantime, it had "entered the history of cinema and costume, [the title] became a classic saying, an emblem of disproportionate excesses in every field of life and thought."[1] Audience comments remain as divided today as they were at the time of the film's release. Some people consider it a masterpiece, some people think it's disgusting, some people think it's simply boring.[2] Meat, represented in a very realistic way, remains a symbol of excess, not an indication of the killing of a living being.

Even more difficult is to understand the audience's reaction to Tretti's film, a film that, despite the enthusiastic comments of intellectuals such as Federico Fellini and Alberto Moravia (Lanzarotti 2010), has remained restricted to the narrow domain of cineforums.

And what about *Carnage*? *The Independent* titled its review "Simon Amstell has made the world's first vegan comedy that's actually funny."[3] And it went on: "*Carnage* is an almost perfect example of how to push a worthwhile message without being preachy." *Carnage* is available online on the BBC website, so it can reach, at least potentially, a much wider audience than film forums. However, just as *Power* has remained a film for a few intellectuals, *Carnage* risks remaining a film for vegans. Very funny, but only those who are already convinced of the goodness of its message watch it.

More interesting, from the point of view of distribution to the general public, are the movies that feature animals. There is a tradition of children's stories and movies with animals as characters, which somehow carry a pro-vegetarian message. The way the animals are represented sometimes makes their ideological and discursive position ambivalent, but their ability to reach audiences is certainly higher than in the films mentioned above. A novel like *Charlotte's Web* (1952, E.B. White) and movies like *Babe* (1995, Chris Noonan), *Chicken Run* (2000, Peter Lord and Nick Park), and *Okja* (2017, Bong Joon-ho) all have in common anthropomorphized animals as characters that share a common destiny: to be eaten in the future. In *Charlotte's Web* (from which a cartoon was made in 1973 and a film in 2006), the main character is a small pig named Wilbur, who lives happily in a barn. His best friend is the spider Charlotte, who tells him exciting stories. One day, Wilbur discovers that the farmer is fattening him up for Christmas dinner. Thanks to Charlotte's friendship, and a little girl's love, the piglet is eventually saved. *Babe* tells the story of another piglet (Babe) who is also supposed to become Christmas dinner for the family of a farmer. Babe succeeds in winning the farmer's affection and becomes a "sheep-pig" instead (the movie is taken from the novel *The Sheep-Pig*, by Dick King-Smith, 1983). Like Wilbur, Babe is a very cuddly pig and can speak with the other animals on the farm. The film thus appears to have been produced primarily as a children's movie, however, the adult audience and critics praised it so much that it received several Oscar nominations (and grossed $254,134,910 worldwide).[4]

The movie has also been successful in delivering a vegetarian message. In the year after the release, the US Department of Agriculture showed a stagnant demand for pork, as reported by the *Vegetarian Times* (O'Connor 1995, 16), and many people claimed to have become veg*ns on account of it, including the actor who played the farmer, James Cromwell (Nobis 2009). This has resulted in the coining of the term "Babe vegetarians." The story of Babe, who changes his destiny from being an animal to be eaten by assuming the role of a shepherd dog, conforms to the dichotomy, typical of carnism, between edible and inedible animals; if the message is pro-vegetarian, then the overall discourse is still framed within carnism.

Released about 25 years later, *Okja* is the story of another pig. Here, the main character is not a piglet, but a giant and genetically modified animal. For Okja, too, the final destination is the slaughterhouse. Like *Babe*, the movie delivers a clear message. It is more direct, as it deals with industrial agribusiness, forced insemination, the mass slaughter of factory animals, and ecoterrorism. However, the effectiveness of the communication risks limitation by the appearance of the genetically modified pig: it is true that she's cute, but she has more similarities to a manatee or a dog than to a real pig. Again, "Okja is endearing because she mirrors these human-constructed qualities that we desire in our companion animals" (Gunawan 2018, 3). The audience's attachment to Okja is created by the fact that she behaves and looks like a non-edible animal (a dog), not because she deserves to be respected as a pig.

Generally speaking, the use of anthropomorphic characters that over-express similarities between nonhuman animals and humans may rouse criticism (Molloy 2011; Parkinson, Twine, & Griffin 2019). The attribution to other animals of human-like qualities prevents people from appreciating them as such, and the natural world as it currently exists; moreover, it depicts a relationship between humans and nonhuman animals based on a specific set of power relations (Parkinson 2019). Similarly, the birds in *Chicken Run* are anthropomorphized animals that can speak; their depiction is therefore far from realistic and again fails to provide any species specificity. All the same, the film has the merit of dealing with the issue of slaughter and the production of animals as commodities (Molloy 2011), and therefore somehow manages to open a fissure in the usual carnonormativity of the media.

In addition to resistant texts, discontent with carnonormativity can be expressed in texts where the dominant discourse is fissured only to produce cognitive estrangement or laughter. In the realms of science fiction, *The War of the Worlds* (1897, H.G. Wells) raises questions about "speciesism." The narrative places Earthlings in confrontation with an intellectually and technologically superior alien species that feeds on them. Questioning the relationship between species is another way to create an opening in the carnist discourse. "We should remember," says the narrator, "how repulsive our carnivorous habits would seem to an intelligent rabbit."

Sometimes, the crack is opened for humorous purposes. Yet it may spark glimmers of reflection, even if an explicit vegetarian message is unintended. *Popeye the Sailor* (1929–, Elzie Crisler Segar) is a comic strip and an animated cartoon for children whose eponymous character is a sailor who becomes extra strong by eating spinach. He sings "I'm strong to the finish 'cause I eats me spinach." The comic effect is triggered by the contrast between the machismo of the sailor and his vegetable diet. However, seeing the cartoon seems to have led children to eat more spinach and less meat (Sirikulchayanonta et al. 2010). *Three Little Pigs*, an animated short film produced in 1933 by Walt Disney, tells the story of three little pigs under attack from a ravenous wolf. In the house of one of them, Practical Pig, there are pictures of relatives and parents: the painting with the title "Mother" shows a sow with her young. In "Father" we see just a string of sausages. In a third, called Uncle Otto, we see a football. In *Three Little Pigs*, the association between the movie characters, which are strongly anthropomorphized, to "the absent referent" (Adams 1991), and to the destiny of death faced by real pigs, was probably conceived as a sort of background humor. However, it can also operate as a fissure, cracking carnonormativity and helping audiences to connect the cute animals on the screen with those killed to be eaten.

In other cases, the fissure aims at different aspects of anthroparchy (such as patriarchy or consumerism), but it also works in fissuring carnonormativity. In two movies from the mid-twentieth century, *By the Light of the Silvery Moon* (1953, David Butler) and *The Giant* (1956, George Stevens), there is a scene of a (patriarchal) family reunion during the Thanksgiving lunch; in both cases, the element that creates the fissure is the relationship between the children and the turkey, which must be sacrificed for the occasion. In the first case, the turkey (Gregory) is saved but he breaks into the dining room during lunch, making much noise and ruining the celebration. In the second, the children discover that the roast brought to the table is Pedro, the turkey that they have fed and made friends with in the previous days. Again, the lunch is ruined (in this case by their crying). As a side effect, the two scenes offer a glimpse of the unfair destiny of "eatable" animals, questioning the dichotomy, typical of carnism, between animals that can be eaten and pets.

Beyond fiction, the cinema world contributes to a critique of carnism through the life choices of many of its protagonists. During the second decade of the twenty-first century, "veganism has turned from a poorly known vegetarian submovement into a way of life praised by some of the world's top celebrities, business people and politicians" (Jallinoja, Vinnari, and Niva 2019, 157). Many Hollywood actors claim to be vegans, such as Benedict Cumberbatch, Zac Efron, Natalie Portman, Woody Harrelson, and Michelle Pfeiffer. To please them, glamorous events, such as the pre-show dinner at the 77th Golden Globes awards (2020), may be plant-based. Moreover, celebrities sometimes cash in on their visibility to launch appeals against animal exploitation and meat consumption, turning the stage of Hollywood's great events into an animal rights forum (as Joaquin Phoenix did at the Academy Awards ceremony, 2020).

12.2 *The Jungle* and more: investigative journalism and the power of the "cognitive trio"

With the necessary adjustments, Comolli and Narboni's frame (1969) also works for the analysis of the "production of literature (which becomes the commodity 'books')" (Comolly and Narboni 1969, 11), and carnism. In the case of literary texts, one can define as "resistant" texts that directly attack carnism and speciesism and experiment with new forms of writing, to make people think about the unjust suffering of nonhuman animals (for instance, the autobiographies and biographies of nonhuman animals). Texts that use direct and perhaps undercover observation, and more generally the sources used by investigative journalism, are also "resistant." Then there are texts written by great novelists, which raise the problem of animal rights, sometimes directly soliciting the reader's empathy. Lastly, there are "fissure" novels that talk about something else but manage, with just one paragraph, or maybe just a simile, to open a crack in the carnist arguments.

An example of "resistant" text, in terms of both content and form, is *Beautiful Joe: An Autobiography* (1894, Margaret Marshall Saunders). To bring the reader closer to the dog's subjectivity, it uses the autobiography technique. Moreover, it offers numerous openings on the life of pets and farmyard animals. For instance, during a train journey, the following dialogue is reported between the lady traveling with the dog Joe and an "old gentleman": "Think of the cattle on the western plains. Choked with thirst in summer and starved and frozen in winter. Dehorned and goaded on to trains and steamers. Tossed about and wounded and suffering on voyages. Many of them dying and being thrown into the sea. Others landed sick and frightened. Some of them slaughtered on docks and wharves to keep them from dropping dead in their tracks. What kind of food does their flesh make? It's rank poison. Three of my family have died of cancer. I am a vegetarian" (*Beautiful Joe: An Autobiography*, 1894).

The anti-carnist strength of the quotation is mitigated by the "dichotomization" proposed in the following lines, where the same "old gentleman" states: "It's a possible thing to raise healthy stock, treat it kindly, kill it mercifully, eat it decently. When men do that, I, for one, will cease to be a vegetarian" (*Beautiful Joe: An Autobiography*, 1894). The novel was the first text published in Canada to sell over a million copies (Harde 2009). Therefore, it is probable that it attained, at least partly, its goal.

Beautiful Joe is a book based on a true story but was destined predominantly for children, and consequently its denunciation is open but adequate for the public. Much more direct is the message conveyed by texts that mix fiction and investigative journalism, which deserve the credit for disclosing the cruelty of what takes place inside a slaughterhouse. Documented reality about the meat industry and fiction are mixed in Upton Sinclair's *The Jungle*, published first in serial form in 1905 and then as a novel in 1906. The story about a Lithuanian immigrant, Jurgis Rudkus, working in a slaughterhouse drew inspiration from the journalistic investigation Sinclair conducted in the Chicago Stockyard undercover. As its focus is more on the exploitative working conditions of humans than on nonhuman animal suffering, the book is considered a "proletarian novel" (Brewster Folsom 1979) more than a manifesto for animal rights. However, while denouncing the lack of attention paid to the quality of the product and the dramatic exploitative conditions of workers, *The Jungle* also exposes animal suffering. In particular, Sinclair highlights the blameless innocence of animals in the face of death and their unfair "deindividualization" (Joy 2010). Every hog is an individual and should be treated as such: "It was all so very businesslike that one watched it fascinated. It was pork-making by machinery, pork-making by applied mathematics. And yet somehow the most matter-of-fact person could not help thinking of the hogs; they were so innocent; they came so very trustingly; and they were so very human in their protests – and so perfectly within their rights. And it was adding insult to injury as the thing was done here, swinging them up in this cold-blooded, impersonal way, without pretense at apology, without the homage of a tear… Was it permitted to believe that there was nowhere upon the earth, or above the earth, a heaven for hogs, where they were requited for all this suffering? Each one of these hogs was a separate creature" (*The Jungle*, 1906).

At the time of its publication, the book was an extraordinary success; it sold more than 100,000 copies within a year[5] and was translated into 35 different languages. The sensation it raised was related more to the representation of a food industry where hygiene was deplorable than to the issues that were close to the author's heart, namely the conditions of workers and the unfair fate of animals. As a result, the public outrage caused by the book stimulated new food quality standards in the US and led to the Pure Food and Drug Act of 1906, but it did not promote a movement against meat consumption.

Investigative journalism has contributed, also in more recent times, to the focusing of attention on the meat industry and its practices. Books and documentaries have multiplied in number, offering increasingly disturbing images and more detailed information. One of the most critical texts is *Beyond Beef: The Rise And Fall of the Cattle Culture* by Jeremy Rifkin (1992), which gives a clear picture from a historical perspective of the role of the meat industry in the United States' regional geography. More informative is *Fast Food Nation* (2001) by the investigative journalist Eric Schlosser. The subtitle of the first edition, *The Dark Side of the All-American Meal*, spells out what the book is about. It tells why Americans have come to eat so much fast food and how the system provides them with such large quantities for little money (Pearce 2011, 457). It focuses mainly on the industrial transformation of fast food and the economy and geography of the ingredients; it describes the transformation of ranch farming, and of small cities following the opening of meatpacking plants and analyzes the conflict between meatpacking companies and traditional ranchers. It even dedicates a few pages, clearly elegiac in style, to the poetics of traditional ranching and the description of one of its protagonists, Hank, who is "handsome enough to be a Hollywood cowboy, tall and rugged, wearing blue jeans, old boots, and a big white hat," but commits suicide at the age of 43 because of the "enormous pressure" of the conflict with industrial farming. A few pages later, the book offers a realistic account of what happens in a modern slaughterhouse. So, like *The Jungle*, *Fast Food Nation* is not explicitly focused on animal advocacy but helps to make the reader aware of the fate of animals. *The Jungle* succeeds in breaking one of the cornerstones of the carnist discourse (deindividualization), by focusing on the individuality of each hog. *Fast Food Nation*, which contrasts the good farmer (Hank) with the evil protagonists of intensive farming, falls into that dichotomization between good meat and bad meat that is one of the pillars of Melanie Joy's "Cognitive Trio"(Joy 2010).

Eating Animals by Jonathan Safran Foer (2009) is also investigative journalism. The book revolves around one question: the author wonders whether it is right that his little son, for whom food seems to be so important, should eat meat. It opens with some reflections on the meaning of the story about what is right to eat and what is not (and questions what justification there is for sparing dogs but eating other animals, for instance, as dogs abound, while other animals have to be bred). One of the book's objectives is to stigmatize the "species barriers": the dichotomy between animals that can be eaten and those that cannot. After, as Upton Sinclair and Eric Schlosser did before him, Foer uses the experience he gathered through an unauthorized visit to an intensive farm to attack his main goal: factory farming, the appalling conditions of the animals, the danger of the abuse of antibiotics, and the consequences for the environment. At this point, however, he falls into the trap of dichotomization: he presents the "good butcher," who butchers, one by one, pigs from non-intensive farms, and the "good breeder," who leaves the animals free to graze in an almost idyllic situation when compared to factory farming. The division between animals that are "good to eat" (because they were farmed correctly and humanely slaughtered) and those that are not to be eaten (because they were bred using intensive farming) makes its appearance once more, and the carnist discourse remains virtually unscathed.

12.3 Consider the animals: empathy and the role of literature

If investigative journalism, sometimes mixed with fiction, can become an instrument of resistance, explicitly denouncing what happens on industrial farms and in slaughterhouses, what can literature do? Can a novel be "resistant"? A similar question about the role of literature is asked in *The Lives of Animals* (1999) by the South African novelist (and Nobel Prize winner) J.M. Coetzee. The book consists of two short stories ("The Philosophers and the Animals" and "The Poets and the Animals"), accompanied by comments from leading thinkers, including the philosopher Peter Singer. Coetzee wrote it, when, invited as a speaker at the 1997 Tanner Lectures at Princeton University, he decided to read two short stories of fiction instead of lecturing conventionally. The two stories present a sort of nesting pattern, as "a metatextual fiction" (Garber 1999, 76). The main character, the Australian feminist Elizabeth Costello, is a writer invited to lecture by an American University (Appleton College) and she decides to speak about animals.

Events are narrated by Elizabeth's son, who happens to be a lecturer at the same university. Elizabeth could have chosen any subject, he says, but, instead of speaking "about herself and her fiction, as her sponsors would no doubt like" (Coetzee 1999, 16), she gives two lectures on vegetarianism and animal rights. In her first lecture, Costello links the "denial" (Masson 2009) that many inhabitants of Treblinka had about what happened inside the concentration camps with the general attitude towards the horrors of the slaughterhouses: we know it, but we prefer not to know it. She also explains that human beings must not place themselves above animals simply because they believe that they are more rational than them; they should instead strive to put themselves in their place, feeling empathy because those who lived around extermination camps also lacked empathy. The reading, and the conversation in the following dinner, also deals with "speciesism" and the distinction made in many cultures, between animals that are good to eat and animals that are not eaten, i.e., dichotomization. The second lecture is organized as a debate about the rights of animals and goes into many issues related to animal welfare policy ("another Western crusade against the practices of the rest of the world," says the philosopher who disputes with Elizabeth) (Coetzee 1999, 60). Before and after the debate, Elizabeth's son and his wife Norma argue about the meaning of veg*n dietary choices ("It's nothing but food-faddism, and food-faddism is always an exercise in power," says Norma, irritated by Elizabeth's vegetarian choice, and her refusal to see any food of animal origin on the table [Coetzee 1999, 67]). The central question is perhaps about the real power of words, literature, and poetry, in representing animals and their sufferings, and it is formulated by Elizabeth's son: "Do you really believe, Mother, that poetry classes are going to close down the slaughterhouses?" (Coetzee 1999, 58). The answer is no, but all the same, Elizabeth does not "want to sit silent" (ibid., 59). The story ends with a final conversation between Elizabeth and her son, and one last question, which touches on one more fundamental point in carnism as a discourse: "group-thinking" (Shaw 2019) or the role played by "the powerful individual whom we call Everybody," as Élisée Reclus wrote in 1901. So, Elizabeth says: "Is it possible, I ask myself, that all of them are participants in a crime of stupefying proportions? Am I fantasizing it all? I must be mad! Yet every day I see the evidences. The very people I suspect produce the evidence, exhibit it, offer it to me. Corpses. Fragments of corpses that they have bought for money." And she adds: "Yet I'm not dreaming. I look into your eyes, into Norma's, into the children's, and I see only kindness, human kindness. Calm down, I tell myself, you are making a mountain out of a molehill. This is life. Everyone else comes to terms with it, why can't you? Why can't you?"[6]

By raising these questions, and answering or not answering them through Elizabeth's voice, Coetzee creates a "fictional device [that] enables him to distance himself from them" (Singer 1999, 91), thus proving his "confusion" (Pick 2011, 9) and the difficulty of rationalizing on such complex issues. Though it raises some points fundamental to carnism as a discourse, Coetzee's contribution has been considered by some critics to be too tied to the human standpoint and unable to present animals as protagonists (Barrett 2014). By putting together "aesthetics and ethics" (Gutmann 1999, 3–4), however, Coetzee provides empirical and philosophical arguments on the ethical question about how humans are supposed to behave towards nonhuman animals; he also succeeds in dispelling many of the accusations made against those who defend the rights of animals and highlights the ecofeminist connections between the various forms of oppression in the world (Wright 2018). At the same time, he advances some fundamental issues about the power of the arts, and in particular the role of literature which is based on "words," something that distinguishes humans from animals, with regard to the representation of animals; so the central question of the text is: "Can literature find a way to get beyond itself in order to respond ethically to the animal?" (McKay 2010, 69).

Being unable to share the thoughts of "Everybody," and also of God, at least according to the interpretation offered by his religion, is the central point of a story by another Nobel Prize-winning writer, Isaac Bashevis Singer. The short story, called *The Slaughterer*, was published in *The New Yorker* in 1967. It is about Yoineh Heir, a man appointed against his will as "the ritual slaughterer" of his religious community. Being very religious, he obeys, in the knowledge that "A man may not be more compassionate than the Source of all compassion." Nonetheless, each killing causes him intense pain, as if he were cutting his own throat. Then, he starts getting uncomfortable with bodies — his own, and other people's. In the month of Elul, when the sacrifices become innumerable and "everybody offers a sacrificial fowl," he dreams of killing a calf, which becomes a girl who asks him for help. He becomes "obsessed" with all living things, all insects, which he loves with an "unfamiliar love," even flies, even mice. Furthermore, he loves human beings less and less, including his daughters, who overeat (meat) and become fat. In his progressive detachment from "Everybody," he detaches himself from God too, who, in the end, no longer seems so merciful to him. He transfigures the landscape and sees blood flowing from the sky, and parts of animals hanging from trees. Eventually, in despair, he commits suicide.

Isaac B. Singer was openly vegetarian and thus committed to the cause of animals. In another short story, he wrote that "for the animals it is an eternal Treblinka" (*The Letter Writer*, 1967). Nonetheless, literary critics often marginalize his vegetarian philosophy (Savvas 2018). Moreover, in *The Slaughterer*, as in *The Lives of Animals*, the narrative focus is on the human protagonist and his emotions, not on animals (Barrett 2014), and thus Singer's story has been analyzed for its ethical-religious content (see Zatlin 1969) more than as advocacy of animal rights. The theme of empathy, which the slaughterer demonstrates through his ability to identify with the pain of "others," is nevertheless its main element. Moreover, the story presents a crescendo. Violence against animals is superimposed first on the main character (who suffers as if the knife cut him), then on other human beings, and eventually on everything around the slaughterer, even the trees and the river. The final images of dead animals and their dismembered bodies on the trees, the blood-stained river, and the sky, display compassion for animal suffering not only through the emotions of a single human being but the whole world.

David Foster Wallace takes a step further along the path of empathy, albeit with a very different style, in *Consider the Lobster*. The essay, written in 2004, is now considered a crucial contribution to posthumanism (Giles 2007; Kaiser 2014). As in Coetzee's case, for Wallace writing the article was the result of a specific circumstance, namely an invitation to participate at a food festival and to write a review of it. The article (originally published in *Gourmet* magazine, August 2004) was written in answer to a request from the magazine's editorial staff for Wallace to describe his experience at the Maine Lobster Festival of 2003. After describing the festival, he gets to a set of questions: "Is it all right to boil a sentient creature alive just for our gustatory pleasure? A related set of concerns: Is the previous question irksomely PC or sentimental? What does 'all right' even mean in this context? Is it all just a matter of individual choice?" (*Consider the Lobster, 2004, 243*).[7]

Instead of just a review, Wallace wrote an essay, in which he not only questions the ethics of boiling a creature alive to increase the pleasure of the consumers but also tackles the issue of "Everybody," or "group-thinking," as a justifying strategy for carnism (boiling a sentient creature alive if the creature is a lobster is "normal" and goes unquestioned by the festival's organizers and by its general audience). He also deals with the issue of differentiation between animal species, which deserve respect from human beings. A dichotomy divides cattle from lobsters, as at the Maine Lobster Festival it is possible to throw live animals into the boiling water in front of a complacent audience, while it would be unthinkable that at the "Nebraska Beef Festival [...] part of the festivities is watching trucks pull up and the live cattle get driven down the ramp and slaughtered right there" (*Consider the Lobster*, 2004, 247).[8]

On this, he also adds a note, where he explains that the unease about eating "higher animals" is shown up by linguistic euphemisms such as "beef" and "pork," which can help to separate the meat from the living creature (we would say to make the living creature an "absent referent"), while "lobster," "fish" and "chicken" are words used both for the animal and the meat. Then, describing an encounter with a taxi driver called Dick, who explains that lobsters do not have a brain like ours and do not feel pain, Wallace tackles one more strategy of carnism, denial. Lastly, he explores the topic of the subjective dimension of pain by looking into the lobster's "experiential networks" (Kaiser 2014, 54). He demonstrates that, though different from that of mammals, the lobster has "an exquisite tactile sense" and that the diversity of its sensory system does not make the lobster any less capable of feeling pain. In fact, despite the lobster's inability to scream, you can recognize its pain, no matter what subjective pain it is, from its "pathetic clinging to the edge of the pot" (*Consider the Lobster*, 2004, 251).

Alongside "resistant" texts that oppose the dominant discourse frontally, there are other literary texts that deserve to be remembered, because they open a "fissure" against it, perhaps with a single sentence or a small paragraph. For example, to quote texts mentioned in this book, in *The Conduct of Life*, Ralph Waldo Emerson cites the remoteness of the slaughterhouse and the complicity of the meat-eater; in *The Dolphin People* (2007, Torsten Krol), the author brings the pain of the slaughtered horse to the page with violence, comparing its scream to that of his dying mother. *Truismes* (1996, Marie Darrieussecq) opens with a short sentence taken from Knut Hamsun's *Benoni* that describes the moment of a pig's death. In the narrative, the sentence is inserted in the slaughterhouse scene, where the narrator risks being slaughtered by her mother. *Benoni* (1908) itself is a "fissure" novel because it is the story of a fisherman who gets rich and of his love for a girl named Rosa. Although pigs are not the protagonists, there is a highly dynamic scene that describes the "pig killing day" in which the pig to be killed gradually becomes aware of what is happening, first with amazement, then with anger and finally with impotence. "The hog takes turns at grunting a little and stopping to listen for a bit. He blinks his eyes and tries to understand what these humans are saying...Now the hog has started shrieking; he has this rope around his shoulder which keeps him following the dairymaid and her bowl. What's the idea of these humans with this rope? He shrieks with all his might ...Then the hog is seized by the four legs and hoisted on to the sledge. The men's grip from sheer nervousness and excitement is so needlessly adamantine and invincible, and the animal lies completely overpowered on the sledge... Then the knife goes in...At first the hog doesn't notice anything: he lies for a few seconds and thinks a bit. But then he knows he has been killed and squeals out his stifled cries until he can do no more" (*Benoni*, English translation 1925, 161–163).

In the scene, together with the innocence of the hog, who does not understand why he cannot follow the girl who feeds him every day and carries his bowl in her hands, the imbalance of power is made very evident. The animal is progressively "overpowered." It is difficult not to feel empathy for him. When reading a passage like this, one finds perhaps the answer to Elizabeth Costello's (and Coetzee's) question about the power of writing.

12.4 "How can you watch that stuff?"… "I don't know… How can you eat it?"

The power of communication as a means of resistance lies at the heart of *My Year of Meats* (1998), from which the section title is taken. The novel, written by Ruth Ozeki, is somewhat autobiographical as the author, in the role of the main character, is of American and Japanese origin, and started her career as a documentary filmmaker (Gersdorf 2010). This character is Jane Takagi-Little, a young American-Japanese woman who has to shoot a series of documentaries about meat. The series, titled *My American Wife* and intended for Japanese television, is sponsored by a US meat exporter called BEEF EX, which wants to increase its sales of meat to Japan. Each episode is supposed to represent a different American wife cooking a different meat recipe for her family in front of the cameras. According to the show's producer, Joichi "John" Ueno, the families who should represent America for the Japanese public must be all be middle-class and white. Instead, Jane chooses multi-ethnic families and even a couple of vegetarian lesbians. As her relationship with the production team becomes difficult, Jane delves into the American meat production industry, from feedlots to industrial slaughterhouses, and the consequences for the health of the most vulnerable individuals of its most "unconventional" practices. She does a lot of filming, even without authorization. The producer fires her, but using her images, which she believes are much more powerful than words from a communicative point of view, she becomes the producer of very hard-hitting documentaries.

The novel covers many aspects of anthroparchy. The function of meat as a metaphor for a woman's body underlies the whole narrative, and is highlighted by Akiko, the bulimic wife of "John" Ueno (she is too skinny and John, to make the metaphor more explicit, asks her to "put on meat"). Akiko's story develops on the other side of the Pacific in parallel to Jane's, and is a story of violence (Akiko is beaten and raped by her husband), discovery and agency (inspired by what she sees in Jane's documentaries, Akiko flees Japan to reach the United States). Meat is also the symbol of capitalism and the imposition of Western values and forms of consumption on the rest of the world (Chiu 2001). In addition to being a symbol and metaphor of man-woman, and human-animal binarism, meat is also the means by which geographical binarism is linked with them. As Ozeki writes: "Guns, race, meat, and Manifest Destiny all collided in a single explosion of violent, dehumanized activity" (*My Year of Meats*, 1998)

The novel touches on many themes, including ethnicity, mixed ethnicity, hybrid sexuality, the gendered reading of meat consumption (Chiu 2001;

Cheng 2009), the role of the meat industry in environmental transformation and the conquest of the West in the US, food (and cultural) globalization, and environmental justice. It also stigmatizes the role of the media in the promotion of carnism and the feminization of those cooking for their families. Moreover, it opens up a fundamental question about the role of visual documentation in counteracting carnist discourse and, more generally, the role of images in exposing the hidden truth behind appearances.

Despite the strong accusation launched at the US meat industry, the novel does not directly support the veg*n cause. There is no blame for the multi-ethnic family that eats Thanksgiving turkey. Jane herself speaks in a somewhat elegiac tone of her father's family as small ranchers whose business has been swept away by factory farming. It is not even a text where animals take a leading role. On the contrary, they stay in the background, deindividualized, and objectified.

More than a pro-animal novel, *My Year of Meats* is a text that attacks the meat industry and its practices. It is of interest, though, that Ruth Ozeki emphasizes the value of visual evidence. Jane can tell readers what she saw, which happens in the literary form within the novel, but she is obsessed with a visual record. She believes that only by showing what she recorded on video is it possible to weaken the power of the "denial" that helps people to live according to the "collective norm" (that is, according to the power of "Everybody"). Her meat video is "disturbing," as one of the character comments, but also "very important." Hence, the line that gives the title to this section.

12.5 Back to the visual

The question is a recurring one: how is it possible to watch such stuff? More generally, does the video material collected by activists to denounce the meat industry's abominations help the cause, or is it too distressing and, therefore, cannot be seen by the majority of viewers? (Tiplady, Walsh, & Phillips 2013, 2015). The role of the disturbing visual material is evident, not only to the fictional character who asks Jane Takashi-Little how she can watch it, but also to many environmentalists and many human rights activists. A famous saying attributed to Linda McCartney goes, "If slaughterhouses had glass walls, everyone would be a vegetarian."

Visual footage helps to overcome the sometimes invisible but efficient boundaries that surround intensive livestock and meat production facilities (Pick 2016; Marcus 2018). A first film on the subject, silent and now lost, was based on the novel *The Jungle* in 1914. Several visual products representing the slaughter and death of animals were later produced for different purposes. In 1949, the French director Georges Franju directed a short documentary entitled *Le Sang des bêtes* (*Blood of the Beasts*). In black and white, the documentary represents the slaughter of four types of different animals (horses, bulls, calves, and sheep). An emotionless voice comments on what happens, while the killing is "highly systematic, rationalized, and professionalized" (Pick 2011, 136). Interpretation is left entirely to the viewer, who may be horrified by the atrocity of the scenes or fascinated by the aesthetic quality of the images (Sloniowski 1998). Many of the films that later showed images of slaughterhouses owe something to *Le sang des bêtes*' visual approach (Pick 2011): from *Meat* (1976), a short film by Frederik Wiseman, also in black and white, to Richard Linklater's *Fast Food Nation* (2006).

A more comprehensive overview, listing all the uses and abuses of animals by humans in a succession of highly disturbing images, is proposed by *The Animals Film* (1981), by Victor Schonfeld and Myriam Alaux. The movie is described by Pick (2016, 97) as "a suffocating compilation of animals' subjection to violent economies of domination and exploitation." About 25 years later, *Earthlings* (2005, Shaun Monson) adopted the same strategy, when we are once again placed in front of a "graphic exposure of human violence against nonhuman animals" (Pick 2013, 28).

In the new century, the production of documentaries and docu-fiction films related to the topic of animal welfare has become increasingly abundant, thanks to the activism of organizations such as PETA (People for the Ethical Treatment of Animals) and Mercy for Animals, and the commitment of celebrities from the worlds of showbusiness and cinema, such as Paul McCartney, and Joaquin Phoenix. There are films produced by animal welfare organizations, footage shot undercover by animal rights activists, and animal rights documentaries. Besides visual works that document the reality of the meat industry and the cruelty of animal slaughter (*Glass Walls*, 2009; *Dominion*, 2018), there are documentaries and short films about the unfair treatment of animals in farm factories (*Meet Your Meat*, 2002; *Speciesism: The Movie*, 2013), films denouncing the environmental impact of animal-based food products (*Food Inc.*, 2008; *Cowspiracy*, 2014), and also movies that make the effort to demonstrate the benefits of veg★n nutrition and the dangerous consequences of meat-eating (*Peaceable Kingdom*, 2004; *Planeat*, 2010; *Forks Over Knives*, 2011; *Vegucated*, 2011; *Live and Let Live*, 2013; *What the Health*, 2017; *The Game Changer*, 2018). They are distributed using both traditional and new media, increasingly convergent media platforms, and the internet. There are also fiction movies like *Fast Food Nation* (2006), inspired by the book: the film narrates the working life of a fictional character, Don Anderson, the marketing director of a hamburger chain. He travels through Colorado to verify product quality at slaughterhouses and discovers, among other things, the tragic events of the illegal immigrants working there.

Many of these films are just interviews edited together. Others are a compilation of video footage created by industries that exploit animals for internal use, which is then smuggled outside to the public, or is footage made by activists in undercover operations. They work as a "reverse panopticon" (Freeman and Tulloch 2013, 112) because they invert the power of the gaze: thanks to the action of organizations and activists from below, these films reveal what is hidden behind the walls of all the powerful entities that, in one way or another, exploit animals (entertainment activities like circuses and zoos, pharmaceutical laboratories, CAFOs and slaughterhouses).

Earthlings aims to create a biocentric vision of a world to which all animal species belong. Other films concentrate more on human health and veganism and may therefore display a certain degree of anthropocentrism (Pick 2016). Furthermore, with their focus on factory farming, they sometimes fall into the dichotomization between what should not be eaten (intensively farmed animals) and what it is right and fair to eat (traditional extensive farming) (Arcari 2017). Even so, they are all committed to breaking down the hegemonic discourse on meat. Undercover footage, whether available on the internet or edited in a documentary, proves an essential tool for the advocacy of animal welfare (Arcari 2020). Although these films are sometimes a "collage of atrocities" that can upset more sensitive viewers (Tiplady, Walsh and Phillips 2015; Marcus 2018), they can have a significant effect on their audience: according to an online survey conducted in 2013, to which about 7000 people responded, 42% of those who decided to become vegan did so after seeing an educational video or film.[9]

Notes

1 www.lastampa.it/spettacoli/2013/05/13/news/una-grande-abbuffata-1.36093343
2 www.movietickets.com/movie/la-grande-bouffe/89d4494a-1e6c-30f5-8fd0-2782e905ba34/reviews/audience
3 www.independent.co.uk/arts-entertainment/tv/carnage-review-bbc-iplayer-simon-amstell-vegan-comedy-actually-funny-a7636871.html
4 www.boxofficemojo.com/title/tt0112431/?ref_=bo_se_r_1
5 www.encyclopedia.chicagohistory.org/pages/679.html
6 Excerpts from *The Lives of Animals* (1999, 69), John Maxwell Coetzee, Princeton University Press.
7 Excerpt from *Consider the Lobster* by David Foster Wallace, copyright © 2005. Reprinted by permission of Little, Brown and Company, an imprint of Hachette Book Group, Inc. .
8 Excerpt from *Consider the Lobster* by David Foster Wallace, copyright © 2005. Reprinted by permission of Little, Brown and Company, an imprint of Hachette Book Group, Inc.
9 https://faunalytics.org/survey-of-vegans-2013-the-results/#

Bibliography

Adams, Carol J. (1991). *The Sexual Politics of Meat: A Feminist-Vegetarian Critical Theory*. Continuum

Arcari, Paula (2017). Perverse Visibilities? Foregrounding Non-Human Animals in "Ethical" and "Sustainable" Meat Consumption. *The Brock Review* 13(1), 24–53. [https://doi.org/10.26522/br.v13i1.1317]

Arcari, Paula (2020). *Making Sense of 'Food' Animals*. Palgrave Macmillan.

Barrett, Paul (2014). "Animal Tracks in the Margin": Tracing the Absent Referent in Marian Engel's *Bear* and JM Coetzee's *The Lives of Animals*. *ariel: a Review of International English Literature* 45(3), 123–149. https://doi.org/10.1353/ari.2014.0018

Brewster Folsom, Michael (1979). Upton Sinclair's Escape from The Jungle: The Narrative Strategy and Suppressed Conclusion of America's First Proletarian Novel. *Prospects* 4, 237–266. https://doi.org/10.1017/S0361233300002891

Chan, Andrew (2003). "La Grande Bouffe": Cooking Shows as Pornography. *Gastronomica—The Journal of Food and Culture* 3(4), 47–53. https://doi.org/10.1525/gfc.2003.3.4.46

Cheng, Emily (2009). Meat and the Millennium: Transnational Politics of Race and Gender in Ruth Ozeki's My Year of Meats. *Journal of Asian American Studies* 12(2), 191–220. https:// doi:10.1353/jaas.0.0033

Chiu, Monica (2001). Postnational Globalization and (En) Gendered Meat Production in Ruth L. Ozeki's *My Year of Meats*. *Literature Interpretation Theory* 12(1), 99–128. https://doi.org/10.1080/10436920108580283

Coetzee, John M. (1999). The Lives of Animals. In Coetzee, John M., Garber, Marjorie, Singer, Peter, Doniger, Wendy, & Smuts, Barbara (pp. 13–69). *The Lives of Animals*. Princeton University Press.

Comolli, Jean-Luc, & Narboni, Paul (1969). Cinéma/idéologie/critique. *Cahiers du Cinéma* 216,11–16; 217, 7–13 (Browne, Nick (ed.), *Cahiers du Cinéma Volume 3 1969-1972 The Politics of Representation. An anthology from Cahiers du Cinéma nos 210–239, March 1969–June 1972)*. https://monoskop.org/File:Cahiers_du_Cinema_Vol_3_1969-1972_The_Politics_of_Representation.pdf

Cudworth, Erika (2005). *Developing Ecofeminist Theory: The Complexity of Difference*. Palgrave Macmillan.

Freeman, Carrie P. (2014a). Lisa and Phoebe, Lone Vegetarian Icons: At Odds with Television's Carnonormativity. In Macey, Deborah A., Ryan, Kathleen M. & Springer, Noah J. (eds.), *How Television Shapes Our Worldview: Media Representations of Social Trends and Change* (pp. 193–212). Lexington Books.

Freeman, Carrie P. (2014b). *Framing Farming: Communication Strategies for Animal Rights*. Brill.

Freeman, Carrie P., & Tulloch, Scott (2013). Was Blind but Now I See: Animal Liberation Documentaries' Deconstruction of Barriers to Witnessing Injustice. In Pick, Anat, &Narraway, Guinevere (eds.), *Screening Nature: Cinema Beyond the Human* (pp. 110–126). Berghahn Books.

Garber, Marjorie (1999). Reflection On The Lives Of Animals. In Coetzee, John M., Garber, Marjorie, Singer, Peter, Doniger, Wendy, & Smuts, Barbara. *The Lives Of Animals* (pp. 73–84). Princeton University Press.

Gersdorf, Catrin (2010) Authenticity Redux: Ecology and the "Ethics of Representing" in Ruth Ozeki's *My Year of Meat*. In Haselstein, Ulla, Gross, Andrew, & Snyder-Körber, MaryAnn (eds.), *The Pathos of the Authenticity: American Passions of the Real* (pp. 59–73) Universitätsverlag Winter.

Giles, Paul (2007). Sentimental Posthumanism: David Foster Wallace. *Twentieth Century Literature* 53(3), 327–344. https://doi.org/10.1215/0041462X-2007–4003

Gunawan, Michelle (2018). Navigating Human and Non-Human Animal Relations: Okja, Foucault and Animal Welfare Laws. *Alternative Law Journal* 43(4), 263–268 https://doi.org/10.1177/1037969X18802459

Gutmann, Amy (1999). Introduction. In Coetzee, John M., Garber, Marjorie, Singer, Peter, Doniger, Wendy, & Smuts, Barbara, *The Lives Of Animals* (pp. 3–11). Princeton University Press.

Harde, Roxanne (2009). "Better Friends": Marshall Saunders Writing Humane Education and Envisioning Animal Rights. *Jeunesse: Young People, Texts, Cultures* 1(2), 85–108. https://jeunessejournal.ca/index.php/yptc/article/view/37

Jallinoja, Piia, Vinnari, Markus, & Niva, Mari (2019). Veganism and Plant-Based Eating: Analysis of Interplay between Discursive Strategies and Lifestyle Political Consumerism. In Boström, Magnus, Micheletti, Michele, & Oosterveer, Peter (eds.), *The Oxford Handbook of Political Consumerism*. (pp. 157–179). Oxford University Press Oxford.

Joy, Melanie (2010). *Why We Love Dogs, Eat Pigs, and Wear Cows: An Introduction to Carnism*. Red Wheel/Weiser/Conari.

Kaiser, Wilson (2014). David Foster Wallace and the Ethical Challenge of Posthumanism. *Mosaic: a Journal for the Interdisciplinary Study of Literature* 47(3), 53–69. https://doi.org/10.1353/mos.2014.0030

Lanzarotti, Samuele (2010). Il potere. In Galbiati, Alessio, & Rippa, Roberto (eds.), Augusto Tretti o dell'anarchica innocenza di un irregolare del cinema italiano. *Rapporto Confidenziale*. www.rapportoconfidenziale.org/?p=11383

Marcus, Hadas (2018). How Visual Culture can Promote Ethical Dietary Choices. In Linzey, Andrew,& Linzey, Clair (eds.), *Ethical Vegetarianism and Veganism* (pp. 265–275). Routledge.

Masson, Jeffrey Moussaieff (2009). *The Face on Your Plate: The Truth about Food*. W.W. Norton & Company.

McKay, Robert (2010). Metafiction, Vegetarianism, and the Literary Performance of Animal Ethics in JM Coetzee's The Lives of Animals. *Safundi: The Journal of South African and American Studies* 11(1–2), 67–85. https://doi.org/10.1080/17533170903458504

Mills, Brett (2017). *Animals on Television: The Cultural Making of the Non-Human*. Springer.

Molloy, Claire (2011). *Popular Media and Animals*. Palgrave Macmillan.

Newmyer, Stephen T. (1995). Plutarch on the Moral Grounds for Vegetarianism. *The Classical Outlook* 72(2), 41–43. www.jstor.org/stable/i40162291

Nobis, Nathan (2009). The "Babe" Vegetarians: Bioethics, Animal Minds and Moral Methodology. In Shapshay, Sandra (ed.), *Bioethics at the Movies* (pp. 56–73) Johns Hopkins University Press.

O'Connor, Amy (1995). When Pigs Fly. *Vegetarian Times* 220, 16.

Parkinson, Claire (2019). *Animals, Anthropomorphism and Mediated Encounters*. Routledge.

Parkinson, Claire, Twine, Richard, & Griffin, Naomi (2019). *Pathways to Veganism: Exploring Effective Messages in Vegan Transition. Final Report*, Edge Hill University.

Pearce, James I. (2011). A Brave New Jungle: Factory Farming and Advocacy in the Twenty-First Century. *Duke Environmental Law & Policy Forum* 21, 433–467 https://scholarship.law.duke.edu/delpf/*vol21*/iss2/9.

Pick, Anat (2011). *Creaturely Poetics: Animality and Vulnerability in Literature and Film*. Columbia University Press.

Pick, Anat (2013). Three Worlds: Dwelling and Worldhood on Screen. In Pick, Anat, & Narraway, Guinevere (eds.). *Screening Nature: Cinema beyond the Human*. (pp. 21–36). Berghahn Books.

Pick, Anat (2016). Animal Rights Films, Organized Violence, and the Politics of Sight. In Tzioumakis, Yannis,& Molloy, Claire (eds.), *The Routledge Companion to Cinema and Politics*. (pp. 113–124). Routledge.

Rifkin, Jeremy (1992). *Beyond Beef: Rise and Fall of the Cattle Culture*. Dutton.

Savvas, Theophilus (2018). The Other Religion of Isaac Bashevis Singer. *Journal of American Studies* 52(3), 660–68. https://doi.org/10.1017/S0021875817000445

Shaw, Julia (2019). *Evil: The Science Behind Humanity's Dark Side*. Abrams.

Singer, Peter (1999). Reflection On The Lives Of Animals. In Coetzee, John M., Garber, Marjorie, Singer, Peter, Doniger, Wendy, & Smuts, Barbara, *The Lives Of Animals* (pp. 85–92). Princeton University Press.

Sirikulchayanonta, Chutima, Iedsee, Kingkarn, Shuaytong, Poonsook, & Srisorrachatr, Suwat (2010). Using Food Experience, Multimedia and Role Models for Promoting Fruit and Vegetable Consumption in Bangkok Kindergarten Children. *Nutrition & Dietetics* 67(2), 97–101. https://doi.org/10.1111/j.1747-0080.2010.01426.x

Sloniowski, Jeannette (1998). "It Was an Atrocious Film": Georges Franju's Blood of the Beasts. In Leach, Jim, & Grant, Barry K. (eds.), *Documenting the Documentary: Close Readings of Documentary Film and Video* (159–171.). Wayne State University Press.

Stam, Robert, Burgoyne, Robert, & Flitterman-Lewis, Sandy (2005). *New Vocabularies in Film Semiotics*. Routledge.

Tiplady, Catherine M., Walsh, Deborah-Anne B., & Phillips, Clive J.C. (2013). Public Response to Media Coverage of Animal Cruelty. *Journal of Agricultural and Environmental Ethics* 26(4), 869–885. https://doi.org/10.1007/s10806-012-9412-0

Tiplady, Catherine M., Walsh, Deborah-Anne B., & Phillips, Clive, J.C. (2015). Ethical Issues Concerning the Public Viewing of Media Broadcasts of Animal Cruelty. *Journal of Agricultural and Environmental Ethics* 28(4), 635–645. https://doi.org/10.1007/s10806-015-9547-x.

Versteegen, Heinrich (2010). Armchair Epicures: The Proliferation of Food Programmes on British TV. In Gymnich, Marion, & Lennartz, Norbert (eds.), *The Pleasures and Horrors of Eating: The Cultural History of Eating in Anglophone Literature* (pp. 447–464). V&R Unipress.

Wright, Laura (2018). Vegans in the Interregnum: The Cultural Moment of an Enmeshed Theory. In Quinn, Emelia, & Westwood, Benjamin (eds.), *Thinking Veganism in Literature and Culture: Towards a Vegan Theory* (pp. 27–54). Palgrave Macmillan.

Zambetti, Sandro, & Giacci, Vittorio (1973). *Il potere* by Augusto Tretti with an interview with the director. *Cineforum* 128, 881–896.

Zatlin, Linda G. (1969). The Themes of Isaac Bashevis Singer's Short Fiction. *Critique: Studies in Contemporary Fiction* 11(2), 40–46. http://dx.doi.org/10.1080/00111619.1969.10689946

Index

For Product Safety Concerns and Information please contact our EU
representative GPSR@taylorandfrancis.com
Taylor & Francis Verlag GmbH, Kaufingerstraße 24, 80331 München, Germany

www.ingramcontent.com/pod-product-compliance
Ingram Content Group UK Ltd.
Pitfield, Milton Keynes, MK11 3LW, UK
UKHW021307260926
475702UK00048B/850

* 9 7 8 0 3 6 7 2 6 4 9 9 4 *